Smart Computing and Intelligence

This book series aims to establish itself as a medium for the publication of new research and development of innovative paradigms, models, architectures, conceptual underpinnings and practical implementations encompassed within smart computing and intelligence.

The scope of the series includes but is not limited to smart city, smart education, health informatics, smart ecology, data and computational analytics, smart society, smart learning, complex systems-chaos, computational thinking, brain computer interaction, natural/computer interaction, humanoid behaviour, and impact of educational psychology on computing.

The cornerstone of this series' editorial policy is its unwavering commitment to report the latest results from all areas of smart computing and intelligence research, development, and practice. Our mission is to serve the global smart computing and intelligence community by providing a most valuable publication service.

Xuefeng Wei

Artificial Intelligence in Mathematics Education

Cognitive Analysis and Cognitive Simulation

Springer

Xuefeng Wei
College of Education
Ludong University
Yantai, Shandong, China

ISSN 2522-0888 ISSN 2522-0896 (electronic)
Smart Computing and Intelligence
ISBN 978-981-95-1404-5 ISBN 978-981-95-1405-2 (eBook)
https://doi.org/10.1007/978-981-95-1405-2

This research work is supported by the Humanities and Social Science Foundation of Ministry of Education in China "Research on Cognitive Process Simulation, Learning Disability Diagnosis, and Intervention in Elementary Mathematics Problem Solving".

This Springer imprint is published by the registered company Springer Nature Singapore Pte Ltd.
The registered company address is: 152 Beach Road, #21-01/04 Gateway East, Singapore 189721, Singapore

The original version of the book has been revised. The correction to the book can be found at https://doi.org/10.1007/978-981-95-1405-2_11

Contents

Part I
Introduction and Background

Chapter 1
Introduction

1.1 Background

Mathematics is an important part of human culture. With the rapid growth of information and digitization around the world, mathematical literacy has become increasingly important to every citizen and an essential competency that every citizen should possess in modern society. Mathematical problem-solving ability is an important part of mathematical literacy. Mathematics education worldwide has given much attention to student problem-solving ability in the past few decades. In the 1980s, the U.S. programmatic document for school mathematics education, *An Agen*da for Action, explicitly stated that "the mathematics curriculum should be built around problem solving." In September 2011, the Obama administration reauthorized the Elementary and Secondary Education Act (ESEA), which focuses on improving student learning and problem-solving skills (U.S. Department of Education, 1965). Japanese mathematics education has also attached great importance to "problem solving." In 1994, Japanese mathematics education started to implement the new mathematics curriculum in an all-round way and added "teaching with the subject" to the syllabus. The process of "teaching with the subject" is based on the characteristics of "solving problems" in math classes (Zhang & Tang, 2005). The well-known Cockcroft Report (1982) emphasized, "Problem solving should be regarded as an important part of curriculum theory." In China, according to The Compulsory Education Mathematics Curriculum Standards (2011 edition), "mathematics courses should enable everyone to receive a good math education and to improve every student's ability further on the premise of admitting the individual differences between students. Students' capacity to discover and ask questions is the foundation of innovation (Ministry of Education of the People's Republic of China, 2012)." In addition, the 2022 edition (Ministry of Education of the People's Republic of China, 2022) highlights that the compulsory education mathematics curriculum should enable students to develop the key competences needed for social and personal development through the study

X. Wei, *Artificial Intelligence in Mathematics Education*, Smart Computing and Intelligence, https://doi.org/10.1007/978-981-95-1405-2_1

of mathematics. The cultivation of students' ability to solve problems is an essential educational function of mathematical thinking.

Improving students' problem-solving skills is an effective way to reduce their academic burden. The appropriate design of mathematics questions and practical problem-solving strategies can support student thinking processes and knowledge development to enhance their conceptual understanding. Flexibility in mathematical thinking and application is the ultimate goal of problem solving.

1.2 Focus of Study

The cultivation of mathematical problem-solving ability should be based on the study of mathematical cognition. The 2011 and 2022 editions of *The Compulsory Education Mathematics Curriculum Standards* (Ministry of Education of the People's Republic of China, 2022) noted that "the content of the curriculum should respect the rules of students' cognitive development, including not only the results of mathematics but also how the results are formed as well as the mathematical thinking and methods entailed." Furthermore, "teachers should design instruction based on students' cognitive level and previous knowledge." Finally, they emphasize: "The main purpose of the evaluation should not only focus on the learning results but also attach importance to the learning process."

Early research revealed that teachers lacked an understanding of students' cognitive process of mathematics problem solving at the primary school level. The teaching of mathematics is commonly based on teachers' personal experience, which often results in an incorrect diagnosis of the problems experienced by struggling students. Classroom interventions for struggling students have focused primarily on asking students to correct errors by using the same instructional direction that was originally used. This method fails to provide targeted and meaningful interventions, and students have difficulty recognizing and understanding their mistakes.

In recent years, the development of the learning sciences, especially brain science, cognitive psychology, cognitive neuroscience and related fields, has provided a broader basis for expanded research into the cognitive process of solving mathematics problems. The learning sciences involve interdisciplinary research into teaching and learning. The primary objective of this field is to understand the cognitive and socialization processes that result in the most effective learning. The secondary objective is to design effective and innovative learning environments, including school classrooms and informal settings (Sawyer, 2005).

The teaching process depends on the learning process. Problem solving is the core component of mathematical learning. Research on the cognitive process of problem solving and analysis of student mathematical learning can inform the cognitive rules of student mathematical thinking. The aim of this study is to develop a methodology that facilitates the analysis and description of the cognitive process of mathematical problem solving in primary school students.

1.3 Book Structure

Based on the achievements of previous scholarship, this book proposes a method for problem solving via cognitive simulation. The mathematics problem is used as an example to conduct empirical research. In this book we discuss the application of cognitive analysis and simulation in classroom mathematics teaching (such as the cognitive process of elementary mathematics inquiry question design, a "one-on-one" cognitive diagnosis mode and application, cognition simulation of the interaction process in a mathematics classroom, etc.). The book's organizational structure is as follows.

In this chapter: The research background is analyzed, and the problem to be solved by the research is articulated.

Chapter 2: The core concepts of the research are defined. A literature review of the process models of general problem solving, mathematical problem solving, and the cognitive process analysis of mathematical problem solving is presented.

Chapter 3: The key issues to be solved in this study, the research methods adopted, the framework, the theoretical basis and the significance of the research are described to ensure the authenticity of the research questions, the scientificity of the method and the operability of the research process.

Chapter 4: The process of solving typical problems is analyzed in the sections "Numbers and Algebra," "Graphics and Geometry" and "Statistics and Probability." The Polya mathematical problem-solving model is further refined within the framework and construction of a cognitive model for solving primary school mathematical problems. The characteristics and application scope of the model are analyzed, and its educational significance is presented.

Chapter 5: The problem-solving cognitive model established in Chapter Four is constructed to enable an analysis of the problem-solving process with respect to two typical problems of "different denominator addition" and "mode." The tools of the Adaptive Control of Thought-Rational (ACT-R) framework are used to perform a cognitive simulation. The analysis is conducted using six students from the fifth and sixth grades of a primary school. The students are selected to participate in an empirical study on the cognitive simulation results of the two types of questions using the oral report method. The results of this empirical study demonstrated that the results of the cognitive simulation are consistent with those of the oral report.

Chapter 6: Based on Chapters Four and Five, the cognitive process of mathematical problem solving is analyzed and simulated. The analysis proposes the design basis and principle of the mathematical classroom inquiry problem. It then takes the example of knowledge points such as "mode" and "cylinder side area" to design typical inquiry problems, apply them in classroom teaching, and analyze the effects of teaching applications.

Chapter 7: The cognitive process based on mathematical problem solving is analyzed and simulated. This chapter intensively captures a series of cognitive operations and cognitive components in the process of problem solving and explores the "one-on-one" cognitive diagnosis process based on the cognitive model. It examines and identifies students' problem-solving cognitive processes, especially for students with learning difficulties. It provides more detailed and targeted guidance and advice on teaching practices to support and enhance the individual development of students.

Chapter 8: On the basis of the introduction of "one-on-one" cognitive diagnosis in Chapter Seven, Chapter Eight takes the typical understanding of procedural knowledge and declarative knowledge as an example to analyze the basis and process of typical problem design. The oral report method, student questionnaires, teacher interviews, and other methods are used to conduct an in-depth analysis of the cognitive diagnosis process and its results and discuss the implications for teaching mathematics.

Chapter 9: Based on research in the learning sciences, Chapter Nine proposes a method to analyze classroom interaction from the perspective of the cognitive process. It takes the "seventh-grade" mathematics classroom "Looking from Different Directions" as the research object to determine the typical classroom interaction and uses the thinking model to analyze the question-answering process. It then implements the cognitive simulation with the ACT-R framework. Based on an analysis of the learning process, three suggestions for classroom teaching are proposed to help teachers design more effective teaching strategies.

Conclusion: The study's main research results are summarized, the process innovation is identified and assessed, and the shortcomings of the study are discussed.

References

Ministry of Education of the People's Republic of China. (2012). *Curriculum standards of compulsory education* (2011 ed.). Beijing Normal University Press.

Ministry of Education of the People's Republic of China. (2022). *The compulsory education mathematics curriculum standards* (2022 ed.). Beijing Normal University Press.

Sawyer, R. K. (2005). *The Cambridge handbook of the learning sciences.* Cambridge University Press.

U.S. Department of Education. (1965). *Elementary and secondary education act.* Retrieved from http://www.ed.gov/esea

Zhang, T., & Tang, C. (2005). Common features and enlightenment of mathematics textbooks of primary schools in USA, Japan and Germany. *Comparative Education Research*, 1.

Chapter 2
Mathematical Problem Solving—A Historical Perspective

2.1 Core Concepts

The precise definitions of the core concepts used in this study must be established to clarify the study's relevance, effectiveness, and objectives. The core concepts used in the study are problem, problem solving, cognition, cognitive model, cognitive simulation, ACT-R model, and cognitive diagnosis.

2.1.1 Problem

The first step in problem solving is to define the problem, its nature and its characteristics. For this purpose, Gestalt psychology's definition of the problem is widely cited: "When an organism has a goal but does not know how to achieve the goal there is a problem" (Gilhooly, 1988).This definition consists of the following four points:

Knowledge and abilities determine whether a problem exists. For example, performing subtraction or addition involving three carries may be a problem of significant difficulty for a first-grade student but not for a high school student.
A problem no longer exists as a problem if the goal has changed or if there is no longer a desire to solve the problem.
Problems exist only when there is an awareness of differences between the goal and the current situation.
Problems are goal oriented. They are conceived in relation to a specific goal and thus end when the goal is achieved.

Newel and Simon classified problems into three types according to their characteristics: well-structured problems, moderate-structured problems and ill-structured problems (Newell, 1972). Anderson categorized questions that students often encounter as well-structured problems and ill-structured problems (Anderson, 2000).

X. Wei, *Artificial Intelligence in Mathematics Education*, Smart Computing and Intelligence, https://doi.org/10.1007/978-981-95-1405-2_2

The problems examined in this study are well-structured mathematical problems, such as monism equation problems, arithmetic problems, and cylindrical side area problems; these problems require the application of mathematical knowledge.

2.1.2 *Problem Solving*

Cognitive psychologists commonly define problem solving as follows: "Problem solving is defined as a series of cognitive processes that points to the goal." This definition consists of three parts:

(1) Problem solving is goal oriented. It starts with an aim to achieve a specific goal and ends when the goal is reached.
(2) Problem solving consists of a series of operations. The performance of operations typically requires a logical and thoughtful approach.
(3) Problem solving comprises a series of cognitive operations. It is essentially a thinking activity. Problems in this study refer to problems to be solved through a series of cognitive operations and the application of mathematical knowledge.

2.1.3 *Cognition*

Cognition refers to an individual's ability to acquire knowledge and solve problems, that is, the process and ability of information processing. The definition of cognition consists of two parts:

(1) Cognition includes operations and abilities.
(2) Cognition comprises a series of thinking activities that occur from the point in time when an individual reads a problem to the point in time when the individual answers the problem.

2.1.4 *Cognitive Model*

The term cognitive model originated in the field of computer science and refers to the simulation of the mental process of humans during problem solving and psychological task processing. In many cognitive psychology studies (Anderson et al., 2003; Baddeley & Logie, 1999; Ericsson and Simon, 1993; Healy, 2005; Kalchman et al., 2001; Newell & Simon, 1972; Siegler, 2005), this term has been used to simplify the problem-solving process of human beings and is often likened to a computational model that reflects the human cognitive process. Research shows that cognitive models have effectively predicted and explained the information processing procedures for many problem-solving behaviors (Ericsson & Simon, 1993). The definitions of a cognitive model can be summarized as follows:

(1) A cognitive model is an abstraction and generalization of the cognitive process that actually occurs.
(2) A cognitive model can effectively predict and explain problem-solving behaviors. In this book, the term cognitive model refers to the cognitive analysis framework.

2.1.5 Cognitive Simulation

Many researchers have applied a computational simulation approach to understand the mental process of human problem solving. Based on the cognitive model, problem-solving cognitive simulation first analyzes the cognitive process of problem solving and then programs the cognitive process sequences. It then simulates the process with computer software to make the process of problem solving visible. Problem-solving cognitive simulation assists in the understanding of complex cognitive processes. A limitation of cognitive simulation is that student motivation, feelings, emotions, attitudes and other factors are not considered.

Although there are still many shortcomings of current computer simulations of problem solving, the computer programs are logical, consistent, and reliable. Computer simulation makes a significant and essential contribution to the articulation of the problem-solving process that is not possible through other methods. Computer simulation combines some factors in the problem-solving process, reconstructs this process, improves on the previous analysis of experimental psychology, and opens up a path for understanding the problem-solving process as a whole (Wang & Wang, 1992). With the continuous development of artificial intelligence and brain science, computer simulation will play a greater role in the study of problem-solving cognitive processes. The visualization of simulation results will provide reliable solutions for problem design, classroom interaction and problem diagnosis.

2.1.6 ACT-R Model

The ACT-R model consists of the declarative knowledge and procedural knowledge required for problem solving, the goal and a series of cognitive operations to achieve the goal. The model uses the Lisp language. In this book, ACT-R refers to the sequences of primary school students' problem solving and the ACT-R program (or sequence).

2.1.7 *Cognitive Diagnosis*

Cognitive diagnosis, also called cognitive assessment, is the diagnosis of one's knowledge structure, cognitive development, and cognitive process (Leighton & Gierl, 2007). The definition of cognitive diagnosis includes the following points:

(1) Cognitive diagnosis is based on cognitive processes.
(2) Cognitive diagnosis considers only cognitive factors. It does not consider other factors, such as motivation, emotions, and beliefs.

2.2 Process Model of General Problem Solving

2.2.1 *Literature Review*

I. Newell and Simon's problem-solving process model

The human and computer problem-solving model proposed by Newell and Simon in their book "*Huma*n Problem Solving" can be described as a model of problem-solving models (Newell & Simon, 1972). It can be used to explain a wider range of research on thinking (Ericsson & Hastie, 1994). This problem-solving model is comparable with the memory model (Atkinson & Shiffrin, 1968). Specifically, it conceptualizes problem solving as a process of narrowing the gap between the initial state and the target state through effective methods. This model contains two phases, "forming the internal representation of the problem" and "narrowing the gap between [the] current [state] and [the] target state." It proposes a "General Problem Solver, GPS" computer program.

The process of problem solving considers two states, the initial state and the goal state, and the combination of these states is the "problem space." Under the premise of the problem space and information processing theory, the general process of problem solving can be divided into the following two stages:

(1) Understand the problem: The problem solver transforms (or translates) the problem into psychological representations that are to be placed in working memory in the form of propositions, images, etc. The representations are stored in the brain in an internal form or represented externally.
(2) Find a solution: The information stored in working memory starts to activate the knowledge stored in long-term memory to enable the extraction of relevant knowledge and the selection of strategies and methods applicable to the problem. When an effective solution to the problem cannot be found, it becomes necessary to revise the original problem representation. If the problem can be successfully solved, then the internal problem representation is sufficient to represent the problem. In this case, the problem representation is further embedded in long-term memory and becomes new knowledge.

II. Dewey's problem-solving process model

In his book *How We Think*, Dewey (1859–1952) detailed the thinking process of problem solving and proposed a five-step solution to a problem (Dewey, 1910): (i) a felt difficulty; (ii) the location and definition of the problem; (iii) the suggestion of a possible solution; (iv) development through the reasoning of the bearings of the suggestion; and (v) further observation and experimentation leading to the acceptance or rejection of the proposed solution. Dewey suggested teaching math through a problem-solving approach across all grades and all courses. Ausubel (1978) noted that in the general description of the stage of the cognitive process of problem solving, no significant improvements to Dewey's description have been made for more than 60 years.

III. Gick's problem-solving process model

Gick et al. proposed that the universal problem-solving process consists of four stages: problem understanding and characterization, seeking answers, solving the problem and evaluating the solution. The specific process is shown in Fig. 2.1 (Chen, 2005).

(1) Problem Understanding and Characterization

Gick et al. argued that the first step of problem solving is to determine what the problem is. This step articulates the problem to be solved and the existing given conditions; the problem space is created based on this information. This process includes identifying relevant information, forming a semantic understanding, developing an overall characterization, and problem classification.

(2) Seeking Answers

If, after the problem is understood and characterized, the problem solver fails to activate a particular schema, then the problem solver will start looking for a

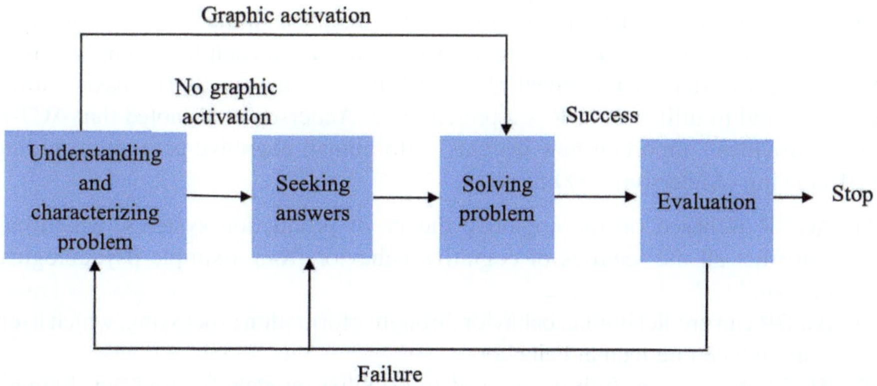

Fig. 2.1 Gick's problem-solving process model

solution. Some commonly used problem-solving strategies are algorithmic strategies, means-purpose analysis, hill climbing, and reverse reasoning.

(3) Solving the Problem

An attempt to develop a solution to the problem corresponds to the process of implementing the solution plan. This step occurs after the problem has been characterized and a problem-solving strategy has been formulated. This process is relatively simple, but it is often overlooked, which leads to errors.

(4) Evaluation

Once a problem answer has been developed, it needs to be evaluated. The evaluation process looks for evidence to confirm or refute the utility and accuracy of the answer. If the answer is proven to be correct, the problem is resolved. If the answer is not correct, the problem-solving process returns to phase 2 or another phase as appropriate and is performed again from that point.

IV. Anderson's ACT-R Model

The ACT-R model was developed and established by John R. Anderson and his colleagues, a group of artificial intelligence educators and psychologists from Carnegie Mellon University. It is a framework to simulate and help explain theories of human cognition. ACT-R researchers have invested substantial effort in understanding how people organize knowledge and produce intelligent behavior. Since 1976 the ACT-R theory has been, evolving from the initial ACT-E model to the current ACT-R model. "R" stands for rationality, which represents "the best way to achieve human goals (Anderson, 1990)." As research has progressed, the ACT-R model has been used to perform a large number of human cognitive tasks through detailed analysis of human perceptions, reflections, and responses to the external environment.

ACT-R is called "a simple theory of learning and cognition." This theory holds that complex cognition consists of knowledge units that are obtained through simple principles. Human cognition is highly complex. This complexity is manifested in the complex combination of basic elements and principles, similar to how computers perform complex tasks through simple operations. To accomplish complex tasks, ACT-R requires that each element of the task must be mastered. The basic knowledge required to utilize ACT-R is a prerequisite. Anderson et al. noted that ACT-R theory provides "important new insights" into human cognitive activity, including the following (Anderson, 2000):

(1) ACT-R is based on the cognitive theory of production systems, which can construct all the features of cognitive behavior from a simple psychological system.
(2) ACT-R can predict human behavior through information processing, which itself can produce intellectual behavior.
(3) ACT-R has successfully been used to establish models for high-level cognitive activities, including scientific reasoning, skills acquisition, and human–computer interaction.

ACT-R theory categorizes knowledge into declarative knowledge and procedural knowledge. Declarative knowledge is "the knowledge of what it is" and refers to the type of knowledge that people know and can express (e.g., China's capital is Beijing, and $1 + 2 = 3$). In ACT-R, declarative knowledge is characterized as small units of primitive knowledge that are called knowledge chunks. Procedural knowledge is "the knowledge of how to do it" and refers to the regular units used to extract declarative knowledge chunks, also known as productions. Squire et al. provide neurological evidence that distinguishes between declarative and procedural knowledge (Squire et al., 1993). Ongoing research into cognitive neuroscience is providing the most up-to-date evidence on more neuronal mechanisms. For example, Graybell found that the basal ganglia are involved in the process of encoding information in the cerebral cortex and are responsible for automating the sequence of actions (Graybell, 1998). One of the consequences of this process is that the sequences that are conscious and slow become automatic and faster.

The ACT-R model is shown in Fig. 2.2. The structure of the model reflects the assumptions of human cognition that are developed based on empirical evidence from psychology experiments. It includes modules, buffers, and pattern matches.

There are two types of modules in ACT-R: perceptual-motor modules and memory modules. The perceptual-motor modules are the interfaces with the real world. The best perceptual movement modules in ACT-R are the visual module and the movement module. ACT-R also includes two memory modules: declarative memory, which

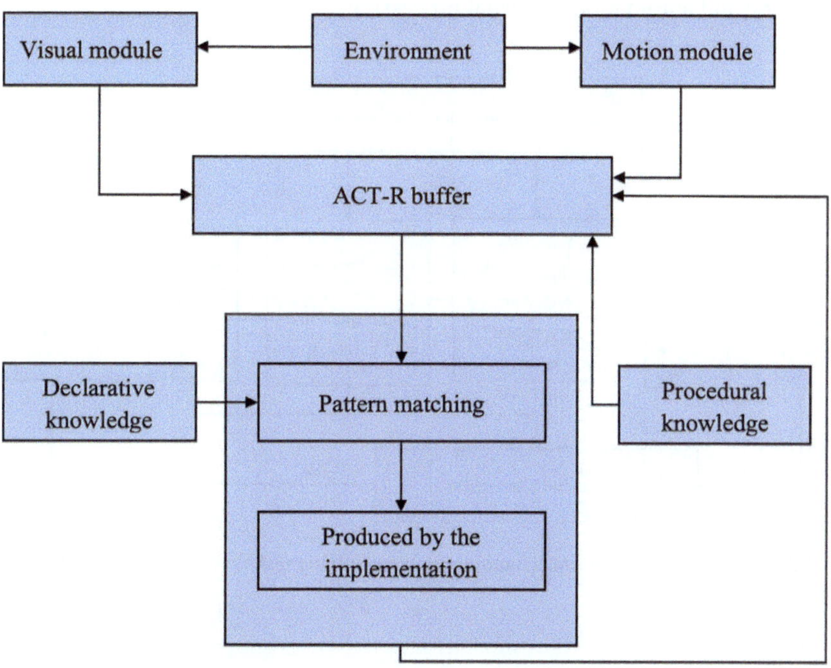

Fig. 2.2 ACT-R cognitive module

contains facts (e.g., Washington is the capital of the United States, $1 + 3 = 4$); and procedural memory, which comprises productions that represent the knowledge of how we do what we do (e.g., how to drive, how to perform addition).

The interaction between the modules in ACT-R (with the exception of the perceptual-motor modules) is performed through buffers. Each module has a dedicated buffer that serves as the module's interface. The contents of the buffer at a particular moment characterize the state of the ACT-R model at that moment.

The pattern matcher looks for productions that match the current state in the buffer. Multiple productions can be matched at a given time, but only one production run is executed. The contents of the buffer are modified during production, changing the state of the system. Hence, cognition in ACT-R is represented as a series of activations of productions.

Many scholars in China have performed a large amount of theoretical and practical research on the cognitive process of problem solving. Wen Gao et al. argued that the general process of problem solving can be attributed to the following five stages: (1) identification and definition of the problem; (2) selection and application of strategies to solve the problem; (3) characterization of the problem; (4) resource distribution; and (5) monitoring and evaluation. Jiang and Yang (2002) divided the problem-solving process into two parts: task understanding and implementation of the operation. Figures 2.3 and 2.4 show the models for task understanding and execution, respectively. The problem-solving information processing mechanism is driven by concepts, with psychological resources flowing in an orderly manner between the processing and automation of partial processing.

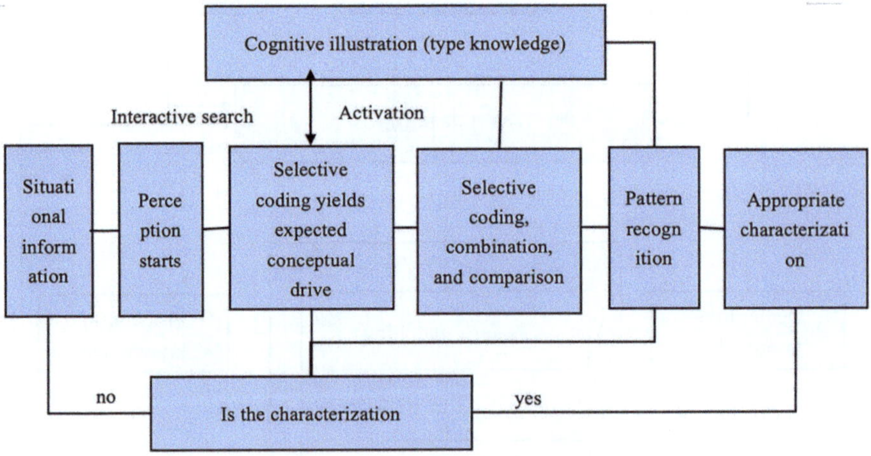

Fig. 2.3 Modular diagram of understanding problem-solving tasks

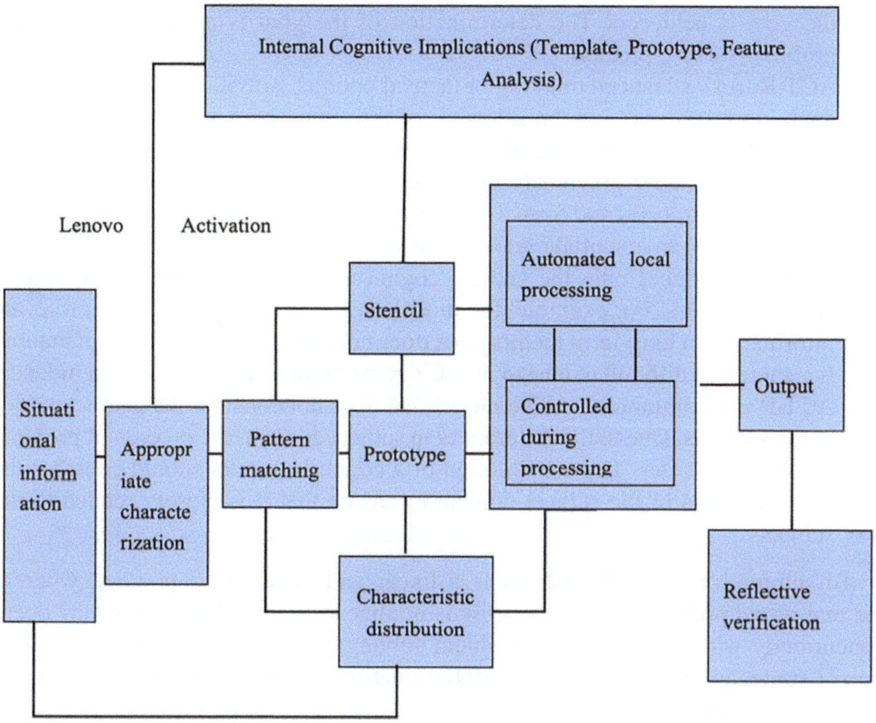

Fig. 2.4 Problem solving in the operation of all modules

2.2.2 Summary

According to Simon, the process of human information processing is a single-linear system that performs a series of activities because a human can only think of and do one thing at a time. People are a single-linear system (Simon, 1978). The proposed problem-solving engineering model provides a precise prediction for certain types of problem-solving behaviors and thinking tests. Hence, human thinking no longer appears mysterious (Newell & Simon, 1961). It is a functional description, is abstract, and is unrelated to the structure. The general problem-solving process and general problem-solving strategy do not consider subject knowledge.

Anderson's ACT-R model provides an abstract cognitive structure that describes the cognitive model from a functional point of view only. ACT-R has been applied in many fields and has achieved some promising results. However, the ACT-R model and problem-solving process are not the same, and many problems still need to be studied.

(1) ACT-R is a serial process of implementing a solution to a problem after the problem has been identified. It does not indicate how the problem is to be determined. The problem is conceived when a goal is identified and ends when

the goal is achieved. The determination of the goal is a crucial step in the problem-solving process.

(2) ACT-R lacks advanced thinking or tactical options. Solving a problem involves applying a strategy, and the selection of the appropriate strategy is the key to successful problem solving. When people solve problems, they usually extract strategies that have previously successfully solved similar problems from their long-term memories and form a new strategy to solve the problem. Formulating the strategy is an essential part of problem solving.

(3) ACT-R introduces only the general cognitive process, without highlighting a specific process. For example, the model mentions the comparison of relevant information in long-term memory but does not provide any further explanation. It is therefore difficult to design an ACT-R program using this limited guidance.

(4) ACT-R is a common cognitive model that does not consider the characteristics of the subjects. Due to the differences in subject content, the process of problem solving will differ. The construction of a cognitive model should consider the characteristics of the subject content to ensure that it is closely related to the subjects.

To summarize, ACT-R analysis is a linear and symbolized process, whereas the real-world problem-solving process is nonlinear. Many other events may occur concurrently with the process of problem solving, such as reflection on the selection of strategies, calculations, and a determination of whether the problem-solving process is complete.

2.3 The Process Model of Mathematical Problem Solving

2.3.1 Literature Review

I. Polya's "table of how to solve the problem"

A famous mathematician and mathematics educator, George Polya (1887–1985) is a landmark figure in the field of mathematical problem-solving research. In his book *How to Solve It*, Polya (2012) proposed four steps of problem solving as shown in Table 2.1.

The four steps of Polya's problem-solving approach have far-reaching implications for mathematics education. The works of today's well-known problem-solving experts in mathematics education, such as Kilpatrick and Schoenfeld, are all based on Polya's work.

II. Schoenfeld's mathematical problem-solving approach

Schoenfeld stressed that four factors should be considered in mathematical problem solving: the knowledge base, problem-solving strategies, self-control, and belief systems (Schoenfeld, 1985). His research revealed that cognitive factors

Table 2.1 Polya's four-step approach to problem solving

(1) Understand the topic	Understand the problem: What are the unknowns? What are the known data? What are the conditions? Is it possible to meet the conditions? Are the conditions sufficient to determine the unknowns? Or are they not enough? Or redundant? Or contradictory? Draw a picture and insert an appropriate symbol Please separate the different parts of the condition
(2) Identify the relationship between known data and unknown data If you cannot find a direct link, you may need to consider secondary topics In the end, you should be able to find a solution to the problem	Proposed solution: Have you seen it before? Or have you seen the same topic appear in a slightly different form? Do you know a related topic? Do you know a theorem that may be useful? Observe the unknowns! Moreover, try to come up with a topic you are familiar with the same or similar unknowns Once you have identified a topic related to your question that has been solved before, answer the following questions: Can you use it? Can you take advantage of its results? Can you take advantage of its methods? To apply it, do you need to introduce any auxiliary elements? Can you recount this topic? Can you describe it in a different way? Revisit the definition If you cannot solve the problem, try to solve a related topic. Can you think of a topic that is easier to solve? A more general topic? A more specific topic? A similar topic? Can you solve part of the problem? Keep only a portion of the conditions, and set the other conditions aside: to what extent can the unknown be known (or defined)? How does it change? Can you derive useful information from the known data? Can you think of other suitable known data to determine the unknowns? Can you change the unknowns or the known data or change both if necessary so that the new unknowns and the new known data are closer to each other? Have you used all the known data? Have you used all the conditions? Did you consider all the key concepts in the title?
(3) Execute your plan	Execute the plan: Run your solution and check every step. Can you confirm that each step is correct? Can you prove that the solution is correct?
(4) Check the obtained answers	Review: Can you test the result? Can you test the argument? Can you deduce the result in different ways? Can you see it at a glance? Can you apply the result or the method to other topics?

play a key role. According to the metacognitive point of view, he divided the problem-solving process into six stages: reading, analyzing, exploring, planning, implementing and testing. Figure 2.5 shows a flowchart of the stages of problem solving.

The explanation of the five stages is as follows:

The first stage is analysis. This stage includes analyzing what the problem means, what the known conditions are, what is required, whether the target appears to be compatible, which major principle or system is relevant or must be observed, and which part of the problem relates to mathematical content, among other considerations.

The second stage is planning. In a sense, planning is a "master control mechanism." Planning is not an independent step; it underlies the whole problem-solving process. Its purpose is to ensure that the activities carried out are beneficial.

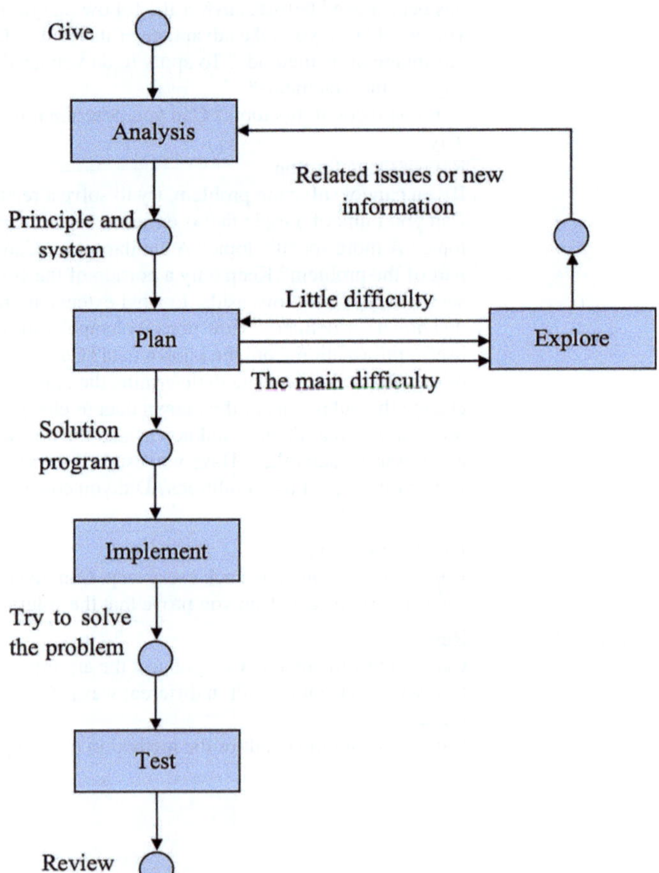

Fig. 2.5 Schoenfeld's problem-solving flow chart

The third stage is exploration. Exploration is the heart of problem solving since the main activities of problem solving are carried out at this stage.

The fourth stage is implementation. It entails the process of implementing a solution to the problem and is the final stage of actually solving the problem.

The fifth stage is testing, a step that should receive more attention. Students seldom check their results; however, checking the results is valuable and important.

To summarize, Schoenfeld's mathematical problem-solving model is based on the work of Polya. His model has been widely recognized in mathematics education.

Regarding the cognitive process of mathematical problem solving, cognitive and mathematical psychologists Lewis & Mayer and Kintsch & Greeno proposed distinct theoretical models along with detailed explanations, which are described below.

III. Lewis and Mayer's mathematical problem-solving model

Lewis and Mayer (1987) noted two important components of mathematical problem solving: problem characterization and solution implementation. Solving mathematical application questions involves characterizing the problem and then applying mathematical or algebraic rules to formulate a solution to the problem. Previous studies (Anand et al. 1987) have shown that the primary difficulty faced by children in solving problems is problem representation rather than calculation. There are two subcomponents of problem characterization (Xin, 2005):

(1) Translation of the question sentences, especially to understand the sentences that indicate the causal relationships. Previous studies noted that it is very difficult to characterize the sentences of causal relationships, and children often ignore or misunderstand causal relationships.
(2) Identification of the type of problem. A child learns to classify problems into various types, that is, to identify a variety of problem-type schemas. If students want to grasp a schema, it is necessary to identify the semantic relations contained in the question. Greeno analyzed the problem-solving process of geometry and math problems and described the knowledge and tactics used to understand and solve a problem. Rily and Heller (1983) noted that as long as children grow, their ability to understand the problem gradually strengthens. These researchers categorized arithmetic application questions into three types according to semantic relations based on conceptual terms such as additions, subtractions, mergers, and comparisons:
(1) Cause-change problem. Cause-change problems describe an increase or decrease in the number of things as a result of additions and subtractions (e.g., "If John has three apples and Jane gives him two apples, how many apples does John have now?").
(2) Combination problem. A combination problem contains a fixed number, and the problem solver needs to perform merging or factoring operations (e.g., "John has three apples, and Jane has two apples. How many apples do they have in total?").
(3) Comparison problem. A comparison problem compares the sizes of two invariable quantities (e.g., "Jane has two apples, and John has three apples more than her. How many apples does John have?").

(3) Green found that students experienced the greatest difficulties in characterizing and solving comparative problems. Lewis et al. (1987) and Verschaffel et al. (1992) found that students encountered more difficulties when the relational words in the comparison were not consistent with the required arithmetic operations.

(4) To explain this difficulty, Lewis and Mayer (1987) constructed a model to compare problem understanding processes. The model focuses mainly on the comparison problem that requires a one-step calculation. These types of problems often begin with an assignment sentence that states the value of a variable (e.g., "John has three apples."). This sentence is usually followed by a causal relation sentence that defines a variable about another variable (e.g., "He has two more apples than Jane does."). Finally, a question about the value of an unknown quantity is posed (e.g., "How many apples does Jane have?").

Lewis and Meyer proposed the hypothesis of consistency, which claims that students show a preference for a specific sentence order. Specifically, students prefer that the sentences be stated in the order consistent with the question. The model shows that when the order in a causal relation sentence does not match the order of students' preferences, more misunderstandings will occur.

IV. According to the model, there is also a particular preference for the presentation forms of comparison problems. When the form of a given causal relation sentence in the question is not consistent with the student's schema, students must rearrange the existing information, and errors may occur during this process. Verschaffel et al. (1992) conducted eye-tracking experiments and revealed that the model is reasonable only when the task performed by the participants has certain cognitive requirements.

The main components of Kintsch and Greeno's mathematical problem-solving model are a set of knowledge structures and a set of strategies for using these knowledge structures in constructing problem representations and performing problem solving (Kintsch & Greeno, 1985). Characterization is twofold: on the one hand, it is a textual frame that characterizes textual input; on the other hand, it is an abstract question representation or problem model that contains question-related information derived from the textual framework.

The model includes two sets of knowledge structures used to characterize and solve problems: (1) a set of propositional frameworks for converting sentences into propositions and (2) a set of schemas that characterize the relationships between features and collections in a general form, which are used to construct macrostructures and problem patterns. After characterization is completed, the problem-solving stage begins.

Other researchers have also investigated the mathematical problem-solving process. Ausubel and Robinson (1969) took geometric problems as prototypes and proposed a problem-solving model. They noted that problem solving generally involves four stages: presenting the problem situation proposition, identifying the problem's objectives and the known conditions, filling the gap, and testing the proposed solution to the question. Mayer argues that the cognitive process of solving an application question can be divided into four phases: characterizing the questions, synthesizing the questions, formulating and adjusting the solution, and implementing the solution (Mayer, 1984).

Starting from the cognitive processing of problem solving, Yu (2008) matched the phases of problem solving with cognitive processing, resulting in the cognitive model of mathematical problem solving shown in Fig. 2.6. He asserted that mathematical problem solving is the process of extracting a problem-solving schema from long-term memory to apply in a new problem situation. He divided mathematical problem solving into four stages: understanding the problem, choosing the operators, applying the operators and evaluating the results. The cognitive processes corresponding to these four stages are problem representation, pattern recognition, problem-solving migration, and problem-solving monitoring.

Zhang and Guan (1997) divided the cognitive process of elementary mathematics application into three phases: characterization, problem solving and thinking summary. Zhu (1999) proposed a "four-step feedback" program for solving mathematical problems. He argued that the psychological process of solving mathematical problems can be divided into four stages: awareness of the existence of

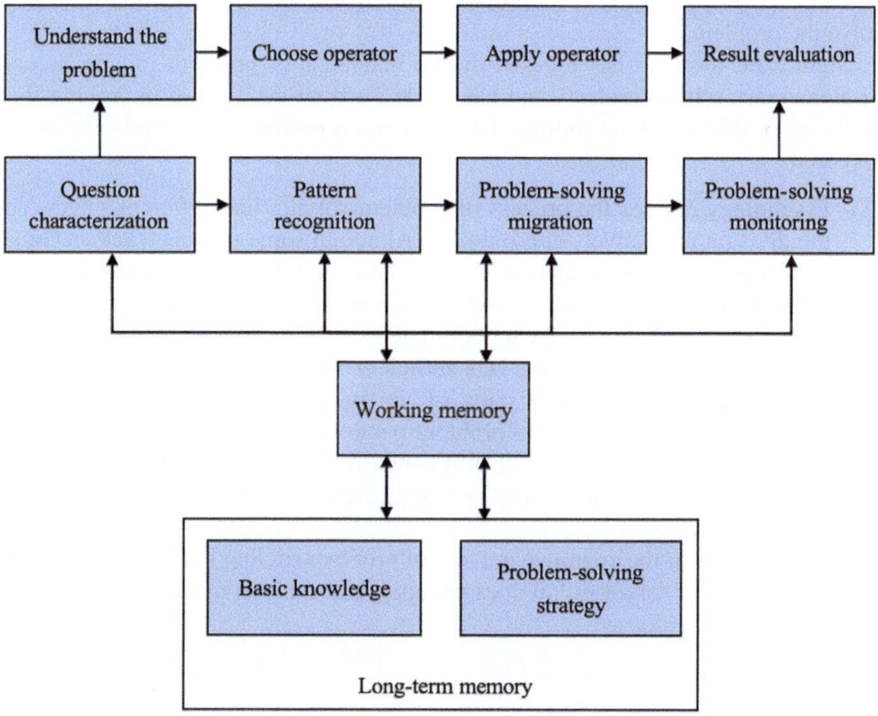

Fig. 2.6 Pin Yu's mathematical problem-solving cognitive model

the problem, characterization of the problem, determination of the problem-solving strategy, evaluation, and reflection (He, 2004).

2.3.2 Summary

Lester noted that a serious consequence of the inadequacies of the Polya model is that meta-cognition is largely ignored by research conducted on the basis of this model. Specifically, this approach focuses only on heuristics. However, the application of algorithms and heuristics depends on complex thinking activities, most of which may be explained by the use of meta-cognition. Many efforts to improve student problem-solving abilities are therefore not successful. The reason is that teaching education overemphasizes the development of heuristic abilities and neglects the regulatory power necessary for adjusting individual behaviors. Although the details of the Lewis model and the Kintsch model are different, both consider that the problem-solving process consists of two main components: problem understanding (or problem characterization) and problem solving (Xin, 2005). In problem understanding, students convert the question text into a semantic representation. Essentially, students reconstruct the problem to solve it. In problem solving, students implement a strategy to solve the problem.

Significant research has focused on the mathematics problem-solving process of elementary school students and has made great strides toward improving their skills and problem-solving abilities. However, many problems still need to be further explored:

(1) Psychology divides the process of problem solving into different stages, and the division is relatively approximate. Although some models (such as Gick's and Yuping's models) analyze the corresponding cognitive processing stages in the process of problem solving, these models do not consider the cognitive characteristics of elementary school students. The analysis and research on each stage of the cognitive process lack sufficient depth.
(2) Psychology has researched parts of the problem-solving process, such as problem characterization and problem patterns. However, it does not address the entire cognitive process of problem solving. Therefore, researchers need to conduct comprehensive analyses and research on the complete process of problem solving.
(3) The analysis of the cognitive process of problem solving is conducted only for the purpose of "analysis"; it does not apply the analysis to teaching.

2.4 Analysis of the Cognitive Process of Mathematical Problem Solving

2.4.1 Representation in the Process of Problem Solving

Representation is a central aspect of problem solving. If a problem is correctly represented, it can be said that half of it has been solved (Simon, 1986). Wertheimer (1985) noted that the typical approach to problem solving involves generating a reasonable problem representation; that is, the problem should be properly organized. Representation is a critical step of problem solving. An appropriate characterization should satisfy the following three conditions: (1) the representation corresponds to the real structure of the problem; (2) the problem components in the representation are properly combined; and (3) the representation involves the application of problem solvers' other knowledge. The results of Kaplan and Simon (1990)'s research show that the insight of problem solving is the result of appropriate problem representation and that suitable representations can be obtained by identifying strong constraints to guide the search and thus make the search highly effective. The characteristics and knowledge of the fields are the main sources of strong constraints, and they can guide the subjects to generate specific and effective problem representations.

Kintsch and Greeno (1985) suggested that the key to solving a mathematics problem is the representation of the problem. The representation of a problem is twofold. On the one hand, it represents the textual input and the proposition of the textual frame (propositional text base). In arithmetic problems, the basic propositional text frame is the relationship between sets. Problem solvers must translate linguistic inputs into such a text framework. On the other hand, it is an abstract problem representation or problem model (problem model) that contains text frames, questions, information, and letters from problem solvers in the field of arithmetic problems. The problem model includes three sets of knowledge structures used to characterize and solve the problem: (一) is used in the problem model to translate a sentence into a set of propositional frames (propositional frames); (二) is used as a general set of schemata (schemata) of the relation between the formal characterization and the set; and (三) is used as the general form to represent a set of action diagrams (action diagrams) of computational and arithmetic operations. The problem solver should infer the information needed in the text but not in the text frame when they construct problem models and should exclude any unnecessary information for solving the problem within the framework. After the problem is correctly characterized, the problem solver can start solving the problem. At this stage, they may need to apply problem-solving programs (or sequences).

Ashcraft (1992, 1995), Campbell (1995), Thevenot et al. (2007) found that people directly extract answers from long-term memory when solving simple arithmetical problems. Campbell (2001) and Seyler et al. (2003) found that people adjust their calculation methods according to the operation type and tend to use extraction to perform addition and multiplication operations but deduction for subtraction and division operations.

The results of Qinsikaya's experiment proved that the processes of answering application questions and other questions are the same. Namely, they entail analysis and synthesis (Qinsikaya, 1962). The results of experiments by Zhu and Bai (1964) revealed that students could not answer application questions simply via the process analysis method or the comprehensive method, which is considerably more complex than the application of these two methods. The cognitive process of students' answers to compound applications can be divided into three stages: grasping the subject directly in relation to the things, revealing hidden things, and testing hidden things. Among them, revealing hidden things is the basic stage of solving application questions, and four types of intelligence operation are explored by abstracting words, replacing images, demonstrating activities, changing conditions and performing practical operations.

Zhu (1983) studied the function and performance of pattern recognition in solving students' geometry problems. The results indicate that when working on geometric problems, students need to first identify and classify the problem and then recognize the geometric patterns, with the goal of solving the problem effectively from the given problem situation. Tieru analyzed the verbal outputs of subjects while they were solving problems and found that in solving algebraic equations, these outputs were used mainly for the identification of types of pattern recognition problems. The participants could identify the types of questions that they could quickly and accurately answer and the type of application questions. Moreover, they could distinguish these questions from the need to determine the structural relationship between the semantic context of the specific topic and the general topic, which depends not only on participants' understanding of the current problems of information processing but also on the relevant information stored in the memory search (Shi, 1985). Xie constructed a cognitive process model of the abacus (Xie, 2009). Fu and He used intelligent mathematics questions for experimental homework. They analyzed the results of question representation and problem solving among 34 university students and explored the information processing process during question representation. The results suggest that the information processing process of problem representation can be divided into three stages: searching for and extracting problem information, understanding and internalizing problem information, and discovering metaphorical constraints (Fu & He, 1995).

2.4.2 Representation in the Process of Problem Solving

Studies of the problem-solving process of math application problems show that problem representation can be divided into two stages: sentence representation and structural representation.

Many studies from China and abroad have focused on sentence representation. Zhang (1997) divides the sentences of the application problems into sentences that describe the context, sentences that assign values, question sentences, sentences that describe relations, and compound sentences. The first three sentences are easier to understand, while the other two sentences are the most difficult parts to understand and are key to problem representation. Mayer's research found that the representation of sentences that describe relations is particularly difficult. When the content of the topic is retold by the problem solver, the problem solver often misses the sentences that describe relations or incorrectly describes the relationship feature and even confuses the sentences that describe relations with the sentences that assign values (Mayer, 1987). Lewis and Mayer (1987) noted that sentences that describe relations define a variable based on another variable, and the problem is finding the value of another variable. Their studies on the representation of relation sentences in an experimental setting revealed that it is more difficult to make representations "when the required arithmetic operation is different from the words mentioned in [the] relation sentence." Lewis and Mayer (1987) studied the factors influencing the representation of sentences and found that the overall representation directly affected sentence representation and that students' representation of sentences improved significantly when they received training on representation sentences and overall representation (Leiwis, 1989).

Some scholars have studied comparative sentence representations in depth. Riley and others asked students to listen to application questions and then asked the students to repeat them. Bernardo and others examined the role of symbolic knowledge and the problem-information context (PIC) in the process of transforming relation sentences into mathematical equations through four experiments and showed that relation sentences are the most difficult parts for a student to comprehend in performing problem representation. The researchers advocated specialized training on the representation of relation sentences (Bernardo & Okagaki, 1994).

Among studies of sentence representation factors, Loftus and Suppes (1972) found through eye movement experiments that schema knowledge and linguistic competencies are closely related to sentence representation. Gooney and Swanson (1990) argued that the problem schema is closely related to the representation of the relation and problem sentences. The students with a smaller memory capacity could barely recall relation sentences, and the number of relation sentences that students with a larger memory capacity could recall was even lower.

2.4.3 The Role of Schemas in the Process of Problem Solving

In the process of problem solving, the perception of the problem situation, the understanding of the problem, and the formulation of the problem-solving method are affected by the schema. Bernardo asserted that the problem schema is a combination of principles, concepts, relationships, procedures, rules, operations, and others that are related to the type of problem (Bernardo, 1994).It consists of many aspects: (i) it is an organized knowledge block related to problem solving; (ii) it is a summary and abstraction of successful examples of problem solving; (iii) it can be activated by certain cues in the current problem scenario to predict some unknown cues, which contributes to the formation of problem representation; and (iv) it combines strategies, methods, and procedures of problem solving and even automated operating procedures. Hence, it guides the entire problem-solving process.

Regarding the influence of schemas on the perception of problem situations, Gilhooly argued that correct problem perception is similar to suggesting a problem schema, which implies a direct, prototype-like problem-solving approach (Gilhooly, 1988), and that problem perception is closely related to the schema.

The schema is closely related to problem understanding. Knowledge is a semantic network organized by a number of interconnected nodes. The information provided by the problem activates a node in the semantic network, which in turn activates the relevant network, that is, the relevant schema. The schema can provide information and knowledge to problem solvers and help them understand the problem. Best noted that once the schema knowledge is activated, it guides the problem solver to search the problem space in a specific way and to seek the relevant features of the problem, thus helping to improve the efficiency of problem solving (Best, 2000).

To address different types of mathematical problems, we need to choose an appropriate schema to guide the problem-solving process. Meyer's research on solving geometric application problems shows that the key to solving geometric application problems is finding suitable schemas (Mayer, 1981).

The schema not only has an impact on problem solving but also provides an important basis for students to obtain the solution. Students can use existing schemas when solving familiar math problems. When solving new math problems, students can apply an existing schema to guide the process, modifying the schema continuously during the process to ultimately form a new schema. Therefore, problems and problem solving are mutually influential. Simon noted that once a person or a computer program determines the schema required for the problem and the data required for each schema, these schemas are combined to form a new schema—the problem schema—that indicates each part's relationship (Simon, 1986). Forming a problem schema comprises three processes: excluding surface problems, generalizing, and constructing, while moving the focus from the surface level to a deeper level. First, the process of exclusion entails excluding unimportant details from the surface description, which reduces the amount of information stored. Second, the process of summarization also reduces the stored information while simultaneously transforming it. Third, in the process of construction, there is no longer a reduction but rather an increase in information, including the inference of information that is not directly expressed (Li & Wang, 2000), which is also called information beyond actual presentation. The schema is formed through the active cognitive construction of the subject. The process of forming a schema based on understanding is not easy to forget and facilitates migration.

2.4.4 Problem Representation and Problem-Solving Effects

Regarding the relationship between problem representation and problem-solving effects, Anderson found that different representations of problems can produce different problem-solving effects (Anderson, 1993). Anand (1987) revealed that students' incorrect answers to application questions are caused mainly by incorrect representations of the problem structure rather than by computational factors. Yu (2003) proposed the concept-field, concept-system, proposition-field and proposition-system (CPFS) structural theory of mathematics learning. It is believed that students with excellent CPFS structures can more effectively and correctly represent problems and can solve problems and that students who can reasonably represent problems have better CPFS structures. Li et al. (2002) and analyzed the differences between eugenics and students with learning difficulties among 40 third-grade primary school students. The results revealed the following: (1) The differences between high- and low-problem-solving ability students are significant. The differences are related to the representation strategies that students apply in problem solving. (2) The main error that students make in solving the comparative problem is a conversion error and that their error in the inconsistency problem is greater than that in the consensus problem. (3) There are significant differences between the two groups of students' metacognitive knowledge and monitoring skills. Metacognitive monitoring skills have a significant predictive effect on the ability to solve comparative problems (Li et al., 2002).

2.4.5 Review

Based on the discussion above, problem representation plays an important role in the ability to solve application problems. Although this topic has been deeply studied in the field of psychology, some issues remain to be studied:

(i) Studies on problem representation reveal only the characteristics of the representations of problems made by students and rarely explain the reasons for these characteristics.
(ii) Problem representation is one part of the process of problem solving. It does not reflect the entire cognitive process of problem solving. It is therefore necessary to conduct a comprehensive analysis to understand the entire problem-solving process.

2.5 Summary

To improve the relevance and effectiveness of this study, this chapter first defines the core concepts in the study to clarify its significance. The core concepts include problems, problem solving, cognition, cognitive model, cognitive simulation, ACT-R model, and cognitive diagnosis.

Then, the process model of general problem solving, process model of mathematical problem solving, mathematical problem-solving cognitive process analysis and other aspects of the status of research in China and abroad are analyzed. Through this analysis, we found that a large body of research has focused on the process model of mathematical problem solving and cognitive analysis and has made great contributions. However, many problems remain unknown and need to be further explored.

(1) Psychology has conducted in-depth research on certain aspects of problem solving, such as problem representation and problem schema. It has not revealed the cognitive process of the entire problem-solving process. Thus, more comprehensive analysis and research on the entire process are needed.
(2) Although great progress has been made in the field of computer simulation of mathematical problems, it has been achieved from a machine or computer perspective, which is very different from solving a math problem in a classroom setting. Research has not considered the process of students' problem solving. The methods used to solve these problems are often beyond the scope of knowledge available to students; hence, they are unable to provide help and guidance for teaching.
(3) A research team led by Professor Anderson at Carnegie Mellon University studied the cognitive process of math problem solving. ACT-R theory was proposed to guide the simulation and understanding of human cognition, but it does not provide solutions for analyzing the cognitive process of elementary school students' ability to solve math problems.
(4) The existing research provides an analysis of math problem solving from its own perspective. It fails to synthesize the research results of related disciplines and to conduct interdisciplinary studies.

A review of related works provides insights into some remaining questions and serves as the basis for this study.

References

Anand, P. G. (1987). Using computer-assisted instruction to personalize arithmetic materials for elementary school children. *Journal of Educational Psychology, 79*, 72–78.
Anderson, J. R. (1990). *The adaptive character of thought*. Lawrence Erlbaum Associates.
Anderson, J. R. (1993). Problem solving and learning. *American Psychologist, 48*, 35–44.
Anderson, J. R. (1996). ACT: A simple theory of complex cognition. *American Psychologist, 51*, 355–365.

Anderson, J. R. (2000). *Cognitive psychology and its implications* (5th ed.). Worth Publishers.

Anderson, J. R., & Schunn, C. D. (2000). Implications of the ACT-R learning theory: No magic bullets. In R. Glaser (Ed.), *Advances in instructional psychology* (pp. 1–27). Lawrence Erlbaum Associates.

Anderson, J. R., Qin, Y., Sohn, M. H., Stenger, V. A., & Carter, C. S. (2003). An information processing model of the BOLD response in symbol manipulation tasks. *Psychonomic Bulletin & Review, 10*, 241–261.

Ashcraft, M. H. (1992). Cognitive arithmetic: a review of data and theory. *Cognition, 44*, 75–129.

Ashcraft, M. H. (1995). Cognitive psychology and simple arithmetic: A review and summary of new directions. *Mathematical Cognition, 1*, 3–34.

Atkinson, R. C., & Shiffrin, R. M. (1968). Human memory: A proposed system and its control processes. In K. W. Spence & I. T. Spence (Eds.), *The psychology of learning and motivation* (pp. 97). Academic Press.

Ausubel, D. P. (1978). In defense of advance organizers: A reply to the critics. *Review of Educational Research, 48*, 251–257.

Ausubel, D. P., & Robinson, F. G. (1969). *School learning* (pp. 47). Holt, Rinehart and Winston.

Baddeley, A. D., & Logie, R. H. (1999). Working memory: The multiple-component model. In A. Miyake & P. Shah (Eds.), *Models of working memory: Mechanisms of active maintenance and executive control* (pp. 28–61). Cambridge University Press.

Bernardo, A. B. I. (1994). Problem-specific information and the development of problem-type schemata. *Journal of Experimental Psychology: Learning, Memory and Cognition, 20*, 379–395.

Bernardo, A. B. I., & Okagaki, L. (1994). Roles of symbolic knowledge and problem-information context in solving word problems. *Journal of Educational Psychology, 86*, 212–220.

Best, J. B. (2000). *Cognitive psychology* (X. Huang, Trans.). China Light Industry Press.

Campbell, J. I. D. (1995). Mechanism of simple addition and multiplication: A modified net-work-inter-ference theory and simulation. *Mathematical Cognition, 1*, 121–164.

Campbell, J. I. D., & Xue, Q. (2001). Cognitive arithmetic across cultures. *Journal of Experimental Psychology: General, 130*, 299–315.

Chen, Q., & Liu, R. D. (Eds.). (2005). *Educational psychology* (pp. 125). Higher Education.

Dewey, J. (1910). *How we think* (pp. 50). DC Health.

Ericsson, K. A., & Simon, H. A. (1993). *Protocol analysis: Verbal reports as data* (pp. 57). MIT Press.

Ericsson, K. A., & Hastie, R. (1994). Contemporary approaches to the study of thinking and problem solving. In R. J. Sternberg (Ed.), *Thinking and problem solving* (2nd ed., pp. 41–49). Academic Press.

Fu, X. L., & He, H. D. (1995). A study of the process of problem representation. *Journal of Psychology*, 2.

Gilhooly, K. J. (1988). *Thinking: Directed, undirected and creative* (pp. 75–90). Academic Press.

Gooney, J. B., & Swanson, H. L. (1990). Individual differences in memory for mathematical story problems: Memory span and problem perception. *Journal of Educational Psychology, 82*, 570–577.

Graybell, A. M. (1998). The basal ganglia and chunking of action repertoires. *Neurobiology of Learning and Memory, 70*, 119–136.

Greeno, J. G. (1980). Some examples of cognitive task analysis with instructional implication. In R. E. Snow, P. Federico, & W. E. Montagu (Eds.), *Aptitude, learning and instruction* (pp. 1–21). Lawrence Erlbaum Associates.

Healy, A. (2005). Experimental cognitive psychology and its applications. *Choice Reviews Online, 42*(10), 99–105.

Heller, J., Riley, M. S., & Greeno, J. G. (1983). *Development of children's problem-solving ability in arithmetic* (pp. 153–531). Academic Press.

He, X. Y. (2004). Analysis of the psychological process of solving mathematical problems. *Journal of Mathematical Education.*

Jing, Q., & Yang, S. (2002). Information processing mechanism for problem solving. *Journal of Ningbo University (Education Science Edition).*

Kalchman, M., Moss, J., & Case, R. (2001). Psychological models for the development of mathematics understanding: Rational numbers and functions. In S. M. Carver & D. Klahr (Eds.), *Cognition and instruction: Twenty five years of progress* (pp. 1–38). Lawrence Erlbaum Associates.

Kaplan, C. A., & Simon, H. A. (1990). In search of insight. *Cognitive Psychology, 22*, 374–419.

Kintsch, W., & Greeno, J. G. (1985). Understanding and solving word arithmetic problem. *Psychological Review, 92*, 109–129.

Leighton, J. P., & Gierl, M. (2007). Verbal reports as data for cognitive diagnostic assessment. In J. P. Leighton & M. Gierl (Eds.), *Cognitive diagnostic assessment for education: Theory and applications* (pp. 146–172). Cambridge University Press.

Leiwis, A. B. (1989). Training students to represent arithmetic word problems. *Journal of Educational Psychology, 81*, 521–531.

Lewis, A. B., & Mayer, R. E. (1987). Students' miscomprehension of relational statements in arithmetic word problems. *Journal of Educational Psychology, 79*, 363–371.

Li, X. D., Zhang, X. K., & Wo, J. Z. (2002). The difference between solving the comparative problems of the gifted students and the Underachievers in primary school grade three. *Journal of Psychology*, 4.

Li, X. W., & Wang, Y. (2000). *Teaching strategies* (pp. 67–95). Higher Education Press.

Loftus, E. F., & Suppes, P. (1972). Structural variables that determine the speed of retrieving words from long-term memory. *Journal of Verbal Learning and Verbal Behavior, 11*, 770–777.

Mayer, R. E. (1981). Frequency norms and structural analysis of algebra story problems into families, categories, and templates. *Instructional Science, 10*(2), 135–175.

Mayer, R. E. (1984). Mathematics. In R. F. Dillion & R. J. Sternberg (Eds.), *Cognition and instruction* (pp. 127–154). Academic Press.

Mayer, R. E. (1987). *Educational psychology: A cognitive approach.* Little, Brown.

Newell, A., & Simon, H. A. (1961). Computer simulation of human thinking. *Science, 134*, 2011–2017.

Newell, A., & Simon, H. A. (1972). *Human problem solving* (pp. 9). Prentice Hall.

Pólya, G. (2012). *How to solve it* (H. Tu & C. Feng, Trans.). Shanghai Science and Education Press.

Qinsikaya, S. (1962). *Arithmetic teaching psychology* (J. H. Sun et al., Trans.). People's Education Press.

Schoenfeld, A. H. (1985). *Mathematical problem solving* (pp. 101). Academic Press.

Seyler, D. J., Kirk, E. P., & Ashcraft, M. H. (2003). Elementary subtraction. *Journal of Experimental Psychology: Learning, Memory and Cognition, 29*, 1339–1352.

Simon, H. A. (1978). Information-processing Theory of Human Problem Solving. In W. K. Estes (Ed.), *Human information processing (Handbook of learning and cognitive process)* (pp. 271–295). Lawrence Erlbaum Associates.

Simon, H. A. (1986). *Human cognition-the information processing theory of thinking* (J. Q. Cheng, Trans.). Science Press.

Shi, T. R. (1985). The cognitive model of solving algebra applied problems. *Journal of Psychology*, 3.

Siegler, R. S. (2005). Models of categorization: What are the limits? In L. Gershkoff-Stowe & D. Rakison (Eds.), *Building object categories in developmental time* (pp. 433–439). Lawrence Erlbaum Associates.

Squire, L. R., Knowlton, B., & Musen, G. (1993). The structure and organization of memory. *Annual Review of Psychology, 44*, 453–495.

Steffe, L. P. (2002). *Constructivism in education* (W. Gao & B. Y. Xu, Trans.). East China Normal University Press.

Thevenot, C., Fanget, M., & Fayol, M. (2007). Retrieval or non-retrieval strategies in mental arithmetic? An operand recognition paradigm. *Memory & Cognition, 35*, 1344–1352.

Verschaffel, L., De Corte, E., & Pauwels, A. (1992). Solving compare problems: An eye movement test of Lewis and Mayer's correlation hypothesis. *Journal of Educational Psychology, 84*, 85–94.
Wang, J., & Wang, A. S. (1992). *Cognitive psychology* (pp. 307). Peking University Press.
Wertheimer, M. (1985). A gestalt perspective on computer simulations of cognitive processes. *Computers in Human Behavior, 17*, 248–294.
Xie, A. P. (2009). *Research and discuss the cognitive process of abacus*. Tsinghua University.
Xin, Z. (2005). *Problem solving and knowledge building*. Education Science Press.
Yu, P. (2003). Research on the relationship between individual CPFS structure and mathematical problem representation. *Journal of Mathematics Education*, 3.
Yu, P. (2008). *CPFS theory of mathematical learning psychology* (pp. 150). Guangxi Education Press.
Zhang, Q. L. (1997). *Meta cognition and subject education* (pp. 49–51). Southwestern Normal University Press.
Zhang, Q. L., & Guan, P. (1997). Metacognition analysis of application problems of primary school students. *Psychological Development and Education*, 3.
Zhu, D. Q. (1999). Mathematical problem system construction and solution program. *Chinese Journal of Education.*
Zhu, M. S., & Bai, Z. H. (1964). Pupils' thinking activities in solving compound problems. *Journal of Psychology*, 4.
Zhu, X. M. (1983). The thinking process of solving geometric problems. *Journal of Psychology*, 1.

Chapter 3
Methods and Research Design

3.1 Research Questions

A review of the literature reveals that a large amount of research on problem solving has been conducted, especially in the field of mathematical problems. However, many research questions require further study:

(1) In the field of psychology, the analysis of mathematical problem solving has conducted only a certain part of the process of problem solving in depth but has not considered the entire process.
(2) The existing process model of mathematical problem solving is a general model for all mathematical problems and does not consider the cognitive characteristics and mathematical characteristics of students in different stages.
(3) The research team led by Anderson, a professor at Carnegie Mellon University in the United States, conducted an in-depth and meticulous study on the cognitive process of mathematical problem solving and developed the ACT-R model. However, they did not provide any suggestions on how to formulate a mathematical solution with the ACT-R model.

Therefore, the key questions addressed in this book are as follows: how can the cognitive process of solving mathematical problems in elementary schools be analyzed, and how can an ACT-R model for solving mathematical problems in elementary schools be constructed? The research problem can be decomposed into the following three subproblems:

(1) How can a cognitive model for solving mathematical problems in elementary schools be constructed?
(2) How can the problem-solving process be analyzed based on the cognitive model to build an ACT-R model of mathematical problem solving for elementary school and conduct cognitive simulation?
(3) What is the role of cognitive process analysis in math education?

X. Wei, *Artificial Intelligence in Mathematics Education*, Smart Computing and Intelligence, https://doi.org/10.1007/978-981-95-1405-2_3

This book analyzes the solutions of mathematical problems in terms of the cognitive process. The impact of noncognitive factors such as attitudes, emotions, motivation, and beliefs on problem solving is beyond the scope of this study.

3.2 Research Framework

The research framework contains three parts, which are shown in Fig. 3.1.

3.2.1 Construction of the Cognitive Model of Elementary School Mathematics Problem Solving

A cognitive model is constructed to analyze the elementary mathematics problem-solving process. Elementary school students employ different problem-solving methods than adults do because of differences in the characteristics of their thinking and memory development. The typical problems of elementary school mathematics are taken as the analysis object. Considering the psychological characteristics of elementary school students, combined with cognitive psychology, cognitive neuroscience, brain science, and other research, a cognitive model of elementary school mathematics problem solving is constructed.

3.2.2 Elementary Mathematics Problem-Solving ACT-R Model Construction and Cognitive Simulation

The cognitive model is the basis of cognitive simulation. Cognitive simulation refers to selecting typical problems, using the cognitive model to analyze the cognitive process of problem solving, and using the Lisp programming language to write a cognitive program and simulate it with respect to ACT-R. To verify the validity of the simulation, we selected some students for an oral report experiment and then compared the simulation results with oral report experimental data to determine whether they were consistent. With respect to the problem-solving error, the cognitive model can be used to analyze the cause. The simulation of the problem-solving cognitive process can visualize the "internal process," which helps to identify the characteristics of the problem-solving process.

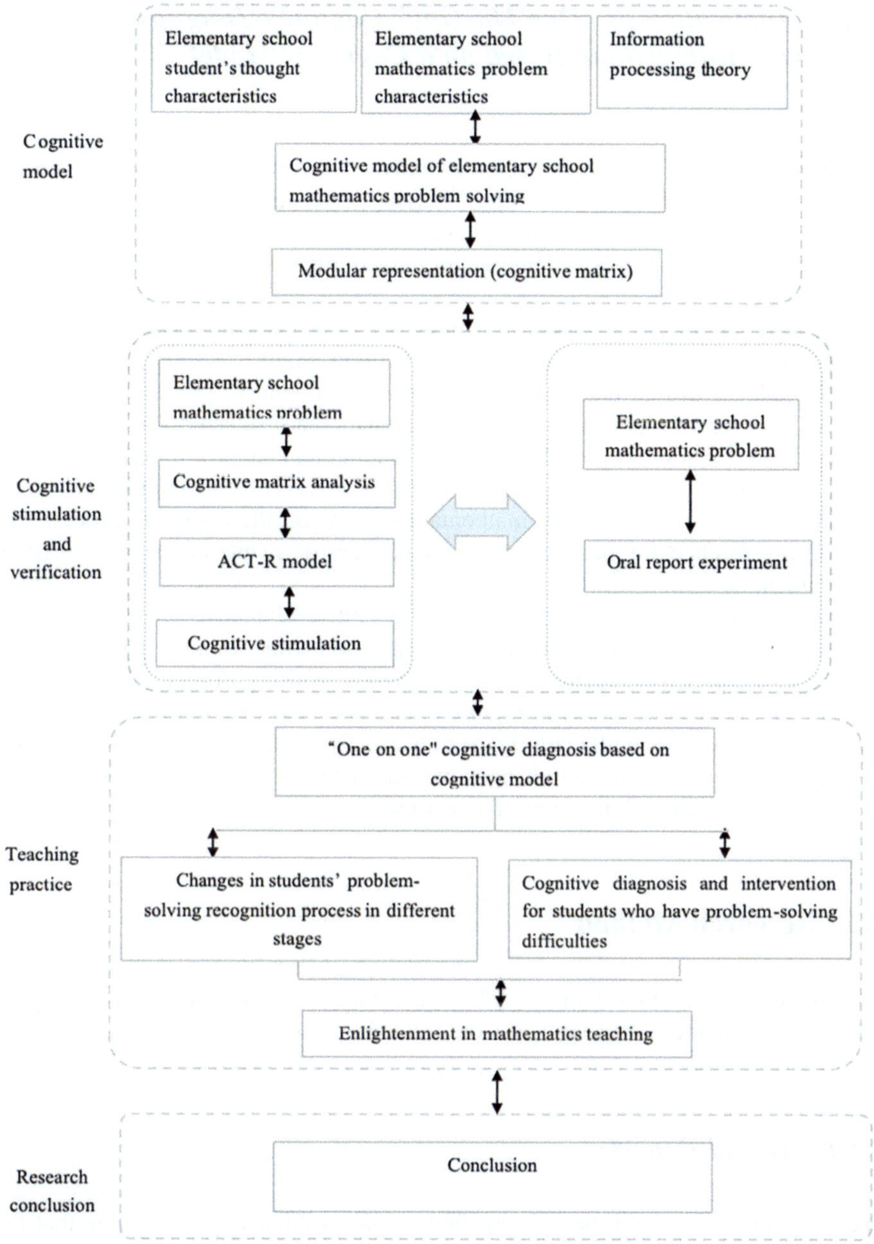

Fig. 3.1 Research framework

3.2.3 Application of "One-On-One" Cognitive Diagnosis and Intervention Based on a Cognitive Model in Teaching

The problem-solving cognitive process is analyzed with respect to solving the practical problems of mathematics teaching. This analysis covers the design of elementary school mathematics problems that are based on cognitive process analysis and "one-on-one" cognitive diagnosis and intervention for students with mathematics learning difficulties.

The design of the elementary school mathematics problem is based on the psychological characteristics and mathematics curriculum analysis of elementary school students and the design basis and principles of the cognitive process. The typical knowledge areas of elementary school mathematics are selected for analysis and design. The typical inquiry of design is applied in elementary school mathematics classroom teaching, and the application effect is analyzed. The analysis of typical inquiry problems of elementary mathematics based on the cognitive process can provide a reference for instructional design, classroom interaction and learning environment design.

This chapter proposes a "one-on-one" cognitive diagnosis and intervention process and method based on cognitive analysis and simulation and then conducts empirical research on students with mathematics learning difficulties. The "one-on-one" method can be used for cognitive diagnosis and intervention for students, especially for students with learning disabilities. This method can assist with mathematics teaching by analyzing changes in both different stages and the same stage of students' problem-solving cognitive process.

3.3 Research Method

Mathematical problem solving is a highly complex activity involving numerous interactive behaviors. This book uses the following research methods.

3.3.1 Verbal Report

Speaking aloud can help many individuals solve problems. It is possible that the auditory memory of what they say helps reduce the burden on working memory. Many problems can be solved in this way, which provides a new method for psychologists to explore to determine how people solve problems.

Verbal reporting is an important method in problem-solving research. It refers to "thinking out loud" thoughts while solving a problem. The researchers recorded the verbal thoughts and analyzed them to reveal the basis of cognitive behavior. Verbal thinking merely expresses the information that already exists in working memory, and it does not affect the process or the outcome of problem solving.

The basic procedure for using the verbal report method is as follows: (1) Before the verbal report is used with the participants, the participants must be trained so that they can solve the problem more smoothly. (2) During the process of verbal reporting, recording equipment is used to record all of the participants' dictations. If there is a pause, the participant should be asked what they are thinking. Unless there is a clear purpose, questions should not be asked because they would interfere with the "think out loud" process. A retrospective verbal reporting approach can be used to ask participants to report specific things. (3) After the verbal report, the researcher classifies and compiles the literary materials verbatim according to the verbal reports and sequences captured by the recording devices. (4) A problem behavior diagram is drawn based on the data analysis, using visualization to present how the student solves the problem.

This book uses the simultaneous verbal report method, which can be divided into four basic steps:

(1) Problem design.

Designing a typical problem based on the purpose of the study is a prerequisite for conducting an oral report experiment. In the study, a typical problem was designed to capture knowledge about concepts such as "mode," "adding fractions with unlike denominators," and "surface area of cylinder," which are used for the verbal report experiment.

(2) Verbal report record.

The pilot's guidelines are highly important. The following is a common example of simultaneous verbal reporting: "Please answer this question. During this process, please speak out loud your thoughts and thinking steps but be mindful not to explain the steps."

(3) Verbal report translation and coding (Liu, 2014).

Eriksson and Simon (1993) noted in "*Protocol Analysis: Verbal Reports as Data*" that the design of the verbal report coding scheme should consider two factors. On the one hand, it must reflect the theoretical idea of the research and meet the theoretical requirements; on the other hand, it must be suitable for the characteristics of the experimental task. Moreover, it should explain the behavior of the subjects during the completion of the experimental task. The corresponding behavioral codes should then be formulated for each statement in the verbal report.

(4) Data statistics and analysis.

By analyzing these verbal protocols, we can infer the process of problem solving. Robertson noted that since verbal reports present the natural situation of solving problems, they can serve as a basis for computer models of problem solving (Robertson, 2004).

During the study, the materials of the verbal report were analyzed according to certain principles, and the information processing was inferred. Simon and Kaplan (1989) noted that the intuitive information usually provided by verbal records concerns the knowledge and information needed to solve the problem rather than the actual processing. Therefore, it is necessary to infer the processing from the information recorded in the verbal protocols instead of trying to directly encode the processing.

3.3.2 Computer Simulation

The computer simulation of problem solving involves writing a computer program based on certain psychological theories to simulate the internal cognitive process of problem solving. This approach enables the computer to solve the problem as human beings would and achieve similar results (Wang & Wang, 1992).

Newell and Simon developed the first computer program to simulate human problem solving, Logic Theorist (LT), and successfully simulated the cognitive process of the human proof of the symbolic theorem. LT proved all 52 theorems in Whitehead's "*Theory of Mathematics,*" which simulates the problem-solving process of human heuristic search. Computer simulation introduced a unique research method to cognitive psychology; since then, computer simulation has become a common method for problem-solving research.

In this book, we used the constructed cognitive model to analyze the cognitive process of solving elementary school mathematics problems. We used the Lisp language to write the cognitive program and implement the simulation in relation to ACT-R. The simulation results were compared with the experimental data of verbal reports to validate the simulation.

3.3.3 Case Study

A case study is a method of research in which detailed consideration is given to the psychological or behavioral development of an individual or a group investigated continuously over a period of time.

During the implementation of the "one-on-one" cognitive diagnosis intervention, we selected the students as representative subjects. We used the verbal report to record the students' knowledge test, preclass exploration, postclass exploration, and first and second cognitive diagnosis interventions. We translated and encoded the verbal report data of different stages for each student and compared and analyzed the changes in students' cognitive processes in different stages.

3.3.4 Interview

An interview is a research method in which the researcher collects the subject's psychological characteristics and behavioral data by conducting verbal conversations with the research subject (Qi, 2004).

For the "one-on-one" cognitive diagnosis intervention research, we designed an interview outline in advance and then interviewed mathematics teachers in the fifth and sixth grades from the target schools to understand the current situation of students' mathematics learning and the common problems that arise in this context. We focused the interviews on the classroom performance and academic grades of students with learning and problem-solving difficulties. The math teacher reported that students with learning and problem-solving difficulties generally have poor grades. Based on this report, we interviewed some Chinese teachers to provide strong evidence for a diagnostic intervention. The entire interview process was recorded and analyzed in a timely manner.

3.3.5 Observation

Observation is a method in which the researcher observes and describes experimental objects and collects research data in a targeted and planned manner through a sensory organ or through the use of certain scientific instruments. The broad definition of observation includes natural observation and experimental observation methods. The narrow definition of observation refers mainly to the natural observation method, which involves examining the observation object under natural conditions. The observation applied in this book is narrow observation. According to different standards, observations can be divided into different categories: direct and indirect; participatory and nonparticipatory; structured and unstructured; and narrative, sampling and evaluation.

We used structural observation in this study. The main test involved designing the content and items for observation in advance and developing the observation form, then strictly following the design and using the form to record observational data. During student activities, the process of students' problem solving was observed with respect to the following questions: Does the student come up with an answer quickly? Does the student take notes? Is the student concentrating during the problem-solving process?

The students' performance during the problem-solving process was recorded via a predesigned "cognitive diagnosis" form. In the follow-up study, the recorded data were coded to obtain quantitative data for further analysis. Structured observation can effectively compensate for the shortcomings of the verbal report method in terms of students' problem-solving behaviors and make the research conclusions more convincing.

3.3.6 Questionnaire

A questionnaire is a research instrument consisting of a series of questions for the purpose of collecting the psychological characteristics and behavioral data of the respondents. The questionnaire method is based on a predesigned survey as the tool, so it has a clear purpose. It can be used to effectively study the various psychological characteristics and behaviors of participants. In this research, to fully understand the status quo of elementary school mathematics problem solving, we developed the "Questionnaire on Elementary School Mathematics Problem Solving for the Current Situation" and the "Outline of Interview on Elementary School Teachers Mathematical Problem Solving for the Current Situation." The reliability and validity of the questionnaire were tested with SPSS 22.0 to ensure its scientificity and objectivity.

3.4 Research Assumptions

This study is based on the following assumptions:

(1) Teaching involves arranging a series of external events and promoting the development of students' internal cognitive processes. The arrangement of teaching activities should be based on the analysis of students' cognitive processes.
(2) Well-structured mathematical problem solving is the main form of elementary school mathematics learning. The analysis and research of the cognitive process of problem solving can help researchers scientifically understand students' cognitive processes and correctly assess students' cognitive rules.
(3) Due to individual differences, each student employs a different way of solving problems. This research focused on the common parts of the problem-solving process.
(4) When the cognitive process of a student while completing a particular task cannot be directly observed and measured, the cognitive process can be indirectly determined by the student's performance upon the completion of the task. A correct answer indicates that the student followed a specific and correct sequence of thinking to determine the answer. This assumption provides a basis for future predictions.
(5) Diagnosis and intervention in the problem-solving cognitive process can help students answer questions correctly and achieve the expected learning outcomes.

3.5 Theoretical Framework

In this book we discuss and analyze problem solving and cognitive simulation using mathematical problems as an example. The theoretical basis includes Piaget's theory of cognitive development, the integration of information processing theory and constructivism.

3.5.1 *Piaget's Theory of Cognitive Development*

Jean Piaget was a Swiss psychologist and the pioneer of the constructivist theory of knowing. Piaget's theory of cognitive development and epistemological view are together called "genetic epistemology." Piaget asserted that the core of the development of children's cognition is the change in the schema. He stated: "Schema refers to the structure or organization of actions. These actions are transferred or summarized in the same or similar environment due to repeated repetition (Piaget, 1980)." He believed that children's cognition from birth to adulthood can be divided into four stages, which are qualitatively different: the sensorimotor stage (from birth to age two), the preoperational stage (from age two to age five), the concrete operational stage (from age five to age eleven), and the formal operational stage (from age eleven to sixteen and above). Each stage has a unique schema, which is associated with different cognitive abilities from those of the previous stage.

One of the characteristics of the concrete operational stage is that its form is not formalized from the content. Piaget called this phenomenon horizontal décalage. Another characteristic of the concrete operational stage is that the resulting system is still incomplete. Zhu and Lin (1986) suggested that the characteristics of children's thinking in this stage gradually transitioned from the concrete image of thinking to abstract logical thinking as the main form, but this kind of abstract logical thinking is still largely associated with the perceptual experience, with a large composition of specific images. They also asserted that this transition is unbalanced in the study of different subjects and materials.

The research subjects of this book are fifth- and sixth-grade elementary school students. Most of them are in the concrete operational stage. The characteristics of children's thinking at this stage can be summarized as follows: (1) Children's thinking is constantly changing in different fields or different materials in relation to the same subject. (2) The construction processes are generally the same, and all follow a common law. (3) Children's thinking is not formalized in this stage and is inseparable from the support of specific items.

3.5.2 *The Integration of Information Processing Theory and Constructivism*

Information processing theory uses a computer analogy to describe the functioning of the human brain and considers that the information processing of the human brain is computable and serially processable. On the one hand, computer technology and intelligence are constantly improving, but they still cannot effectively simulate problems in daily life, such as an "epiphany," which is an irrational and nonlogical problem of human cognition. On the other hand, humans do not follow serial information processing methods to solve problems as computers do. Zhu and Lin (1986) noted that information processing psychology cannot explain human psychology and consciousness, which are the products of human social practice activities, the products of the complete interaction between a subject and an object, or the product of dialectical unity of cognition and emotional will. Therefore, information processing psychology cannot truly explain the social, initiative and creative facets of human psychology (Zhu & Lin, 1986). Although, in terms of problem solving, there are certain levels of similarity between computers and human beings, if human beings are assumed to be computers, that assumption will reflect a "mechanical theory" point of view. A human being is a subjective, initiative, wise, and complex self-organizing system. Currently, computers are capable of handling a large amount of information with some degree of intelligence, but they are not yet fully intelligent.

Resnick noted that when the information processing paradigm was applied to school education, most people accepted the constructivist view (Resnick, 1989). Constructivism includes cognitive constructivism and social constructivism, and cognitive constructivism is derived mainly from the work of Piaget. Kamii and Ewing noted that the three main reasons for the use of cognitive constructivism in education are as follows: (1) it proposes a scientific theory that explains the nature of human knowledge; (2) it proposes a theory that explains children's construction of knowledge theory from birth to adolescence; and (3) it distinguishes three types of knowledge (Kamii & Ewing, 1996). Kamii discovered that children are able to find programs for four arithmetic operations without being taught the common rules (Kamii, 1985; Kamii & Ewing, 1996). This study verified Piaget's theory of the nature of logical mathematics knowledge and effectively demonstrated that the acquisition of children's mathematical knowledge is a process of individual construction. He (1997) noted that the constructivist teaching model emphasizes the student-centered model, regarding the student as the cognitive subject and the active constructor of knowledge. Teachers simply facilitate the meaning construction of students (He, 1997). Social constructivism is derived mainly from the work of Vygotsky, who believed that learning is the process through which individuals construct knowledge and understanding. Social constructivism places greater emphasis on the role of the learner's social, cultural and historical background.

Constructivism is a philosophical concept that does not explain the details of knowledge acquisition or learning, but information processing theory can compensate for this deficiency. In this book, we employ the information processing framework to argue that solving new problems entails a process of knowledge construction. Moreover, the purpose of the application of mathematical knowledge is to better reflect the theoretical integration of the two paradigms of information processing theory and constructivism. This integration does not simply combine basic ideas and theories but rather selects those that are closely related to the research questions as the basic theoretical basis of our research.

3.6 Research Significance

In this book we analyze problem solving and cognitive simulation and conducts an empirical study using mathematical problems as an example. We discuss the application in mathematics classroom teaching and has important theoretical and practical value.

3.6.1 Theoretical Significance

(1) Multidisciplinary knowledge is integrated and a problem-solving cognitive model is constructed to better explain the cognitive patterns of elementary school students. The design of course content should be based on the cognitive rules of students, which cognitive models can effectively reveal. Therefore, constructing a cognitive model is fundamental work that provides the basis and reference for the design of course content.
(2) By examining the process of problem solving, the basis for teaching interventions is determined. Learning outcomes result from the learning process, and problem solving is the main form of school learning. An analysis of the problem-solving cognitive process helps to understand students' learning process and then to design effective teaching methods to achieve the expected learning outcomes.

3.6.2 Practical Value

(1) Understanding, researching and mastering the cognitive process of student problem solving and then incorporating the cognition rules of students in the teaching process can help improve teaching quality.
(2) Teachers' understanding of the cognitive process of student problem solving can help promote the development of students' thinking and innovative ability.
(3) Problem-solving cognitive process analysis can help identify challenges in learning and form cognitive diagnoses for students, especially for students with learning difficulties. The diagnosis results can provide more specific, detailed and targeted guidance and advice for educational practice and individual development.
(4) The study's findings can help teachers learn how to diagnose students' difficulties and problems to improve mathematics classroom teaching and promote the sustainable development of students' mathematical thinking ability.

3.7 Summary

This chapter describes the three key issues to be solved: (1) How can a cognitive model be constructed for solving mathematical problems in elementary schools? (2) How can the cognitive model be used to analyze the problem-solving process, and how can the ACT-R model of elementary mathematical problem solving be built and then used for cognitive simulation? (3) What type of effect can problem-solving cognitive process analysis have on elementary school mathematics teaching?

The main research contents and methodology are determined and analyzed and include verbal reports, computer simulation case studies, interviews, observations, and questionnaires. Five hypotheses for the research process are subsequently proposed. The theoretical basis includes Piaget's theory of children's cognitive development and the integration of information processing theory and constructivism. Finally, the theoretical significance and practical value of the research are discussed.

This chapter clarifies the logical relationship between the research contents, formulates practical solutions, and lays the foundation of the study.

References

Danks, J. H., Ericsson, K. A., & Simon, H. A. (1986). Protocol analysis: Verbal reports as data. *The American Journal of Psychology, 98*(4), 641.

He, K. K. (1997). Constructivist teaching mode, teaching method, and teaching design. *Journal of Beijing Normal University (Social Science Edition)*, 5.

Kamii, C. (1985). *Young children reinvent arithmetic: Implications of Piaget's theory* (pp. 89). Teachers College Press.

Kamii, C., & Ewing, J. K. (1996). Basing technique on Piaget's constructivism. *Childhood Education, 72*(4), 260–264.

Liu, D. Z. (2014). *Study of learning strategy* (pp. 19–22). People's Education Press.

Piaget, J., & Hilde, B. (1980). *Kidology* (W. F. Yuan, Trans.). The Commercial Press.

Qi, D. (2004). *Psychology and education research method* (pp. 161). Beijing Normal University Press.

Resnick, L. B. (1989). Developing mathematical knowledge. *American Psychologist, 44*(2), 162–169.

Robertson, S. L. (2004). *Problem solving psychology* (Q. Zhang, Trans.). China Light Industry Press.

Simon, H. A., & Kaplan, C. A. (1989). Foundation of cognitive science. In M. I. Posner (Ed.), *Foundation of cognitive science* (pp. 1–47). MIT Press/Bradford Books.

Wang, S., & Wang, A. S. (1992). *Cognitive Psychology* (pp. 303). Peking University Press.

Zhu, Z. X., & Lin, C. D. (1986). *Psychology of mind development* (pp. 96–98). Beijing Normal University Press.

Part II
Cognitive Model and Cognitive Simulation

Chapter 4
Cognitive Model for Primary School Mathematical Problem Solving

4.1 Psychological Characteristics of Primary School Children

The Compulsory Education Mathematics Curriculum Standards (2011 edition) notes that the design of the mathematics curriculum in compulsory education fully considers the characteristics of students in a developmental stage, reflecting the students' cognitive rules and psychological characteristics (Ministry of Education of the People's Republic of China, 2012a, 2012b). In addition, the 2022 edition emphasized that the mathematics curriculum should give attention to the teaching level, stimulate the learning interest of students with learning difficulties, encourage them to think positively, cultivate good learning habits, and adapt to the development needs of students (Ministry of Education of the People's Republic of China, 2022).

4.1.1 Characteristics of Primary School Children's Thinking Development

The basic characteristic of primary school children's thinking is that its main form transitions gradually from concrete image thinking to abstract logical thinking. To a great extent, this abstract logical thinking is still directly related to sensual experience and largely still corresponds to a specific image. Piaget also suggested that 7–12-year-old children's thinking occurs in the specific computing stage. Throughout primary school, intuitive teaching is an important method that draws children's attention.

In primary school, children are transitioning from concrete image thinking to abstract logical thinking.

This transition does not happen immediately. It is a complex process (Zhu, 2009)with the following characteristics:

X. Wei, *Artificial Intelligence in Mathematics Education*, Smart Computing and Intelligence, https://doi.org/10.1007/978-981-95-1405-2_4

(1) Children's abstract logical thinking is evolving, but their thought processes still show great specificity.
(2) Children's consciousness of abstract logical thinking is beginning to develop, but they still demonstrate much unconsciousness.
(3) The level of children's abstract logic thinking is constantly improving, and the relationship between the specific image elements and abstract logical components of children's thinking is constantly changing.
(4) The development of children's thinking involves a gradual transformation from a concrete image to abstract logic. It undergoes a clear qualitative change, signifying the critical age at which primary school children's thinking develops.

4.1.2 The Developmental Characteristics of Elementary Children's Internal Language

Piaget was the first to focus on children's self-centered discourse and notice its theoretical significance. However, he does not attach importance to the genesis of egocentric speech or inner speech. Vygotsky (2010) distinguishes between self-centered speech and internal speech. He believes that self-centered speech develops before internal speech does. Both forms of speech have similar functions and similar structures, and self-centered speech disappears in the school-age stage. At this time, inner speech begins to develop. Inner speech is not only an autonomous speech function but also a distinct plane of speech thinking.

In the late period of preschool, children demonstrate the initial germ of inner language but have not developed it enough. Elementary school children are similar to preschool children. In early childhood, children's inner speech is based on the development of language in preschool and gradually develops in the context of school teaching.

Inner speech is not only accompanied by children's activities but also closely linked with children's thinking. Mathematical problem solving is a type of thinking activity. Vygotsky's research on inner speech provides a theoretical and experimental basis for examination and reflection in the process of problem solving. The learning sciences have repeatedly demonstrated the importance of reflection to obtain a deeper understanding of learning. Additionally, the use of brain imaging in the field of cognitive psychology has confirmed the existence of internal circuits in the brain.

4.1.3 Primary School Children's Memory Development Characteristics

Experimental studies have shown that the memory ability of children aged 7–8 years has little difference with that of preschool children (Zhu, 2009). Conscious and abstract logical memorization begin to develop, while unconscious and concrete image memorization understanding still dominate. As children enter the primary stage, conscious and abstract logical memorization and the understanding of memorization gradually become prominent.

In the primary school stage, the task of the teacher is to equip children with sufficient concrete practical material. Teaching from these specific materials continue to develop children's abstract memory of words so that perceptual awareness increases and extends to rational understanding.

Primary school children's knowledge and experience are not rich. They are good at memorizing a specific image. The connection of the first signaling system is most easily established, and the signaling of the second signaling system, which is close to the first signaling system, is also relatively easy to establish. However, the connection of the second signaling system, which is not very close to the first signaling system, is more difficult to establish. On this basis, most of the basic knowledge stipulated in the elementary mathematics curriculum standards is specific knowledge, and some abstract knowledge is closely related to specific knowledge.

With respect to the short-term memory of primary school children, Qian et al. (1989) found that, in terms of the breadth of digital memory, the differences between first and third grades were very significant, but those between third and fifth grades were not significant. Therefore, 7–9 years old marks the rapid development of children's short-term memory capacity. Chen and Wang (2005) found that in the primary stage, the development of memory breadth increases with age.

Compared with short-term memory, working memory places more emphasis on dynamic information storage and processing. Research by Li Deming and others has shown that the working memories of both numbers and language develop with age (or grade) and that the growth rate slows down after the second year of high school (Li et al., 2003).

4.1.4 The Study of Primary School Children's Mathematical Cognition in Cognitive Neuroscience

Cognitive neuroscience emerged in the 1870s seeking to clarify the mechanism of cognitive activity. Cognitive neuroscience studies how the human brain mobilizes the components of each level. The components include molecules, cells, brain tissues, and the whole brain, all of which are involved in performing cognitive activities (Gazzaniga, 1998). Cognitive neuroscience is a new interdisciplinary subject that combines cognitive science and neuroscience. Research on cognitive neuroscience often uses methods such as magnetoencephalography (MEG), positron emission tomography (PET) and functional magnetic resonance imaging (fMRI). Poldrack (2008) notes that these technologies can be used to perform functional brain imaging analysis of brain activity. Poldrack provides reliable evidence about the brain mechanism of cognitive activity to ensure that the research results are scientific.

Many researchers in the field of cognitive neuroscience have studied primary school students' mathematical cognition. Specifically, they have studied the basic processing and brain mechanisms of mathematical cognition. Researchers have attempted to reveal the patterns of brain activity associated with effective mathematics learning.

In recent years, cognitive neuroscience has been accorded great importance for the construction of China: the Chinese *national medium- and long-term plan for science and technology development* (2006–2020) identifies "brain science and cognitive science" as one of the eight leading frontiers of science and technology in the development of science and technology. Great achievements have been made in this field in recent years. With respect to the brain activation mode of addition and multiplication, Zhou et al. (2007) found that the addition operation may rely more on visual–spatial processing, whereas multiplication may be related to language processing. Zhou et al. (2009) compared the electroencephalogram of subjects when they performed multiplication and addition problems. They found that multiplication involves more language processing and that addition involves more activities related to visual imagery processing.

With respect to issues arising with numerical representation in children, Zhou et al. administered the digital Stroop task and found that Chinese kindergarten children (5.85 years old) have a digital automatic processing ability (Zhou et al., 2008). To examine differences in brain function among 8- to 18-year-old children and adolescents on different arithmetic cognition tasks and to address problems in early learning experience and brain plasticity, Zhou et al. (2007) found through an ERP experiment that Continental children presented more negative waves in multiplication problems than did children from Hong Kong. The reason is that children in Hong Kong and Macao perform multiplication differently from Continental children (Zhou et al., 2007).

Qin et al. (2004) investigated the process of solving equations in 11- to 14-year-old children who were learning to solve equations by combining information processing analysis and cognitive neuroscience technology. In this research, the letter coupon and the working model are established. The model includes imaging equation transformation, extracting arithmetical and algebraic knowledge, and arranging the action response. The fMRI method is used to record the course of understanding the equation, which includes the corresponding data on blood oxygen levels (BOLD, blood oxygen level-dependent response) in the frontal, parietal, and motor areas of the brain.

The results of cognitive neuroscience research have shown that the human parietal cortex, especially the area surrounding the bilateral intraparietal sulcus, has a very close relationship with mathematical cognition. Pinel et al. (2001) found that the region was significantly activated when the subjects were comparing the sizes of numbers, and the trend of monotonicity decreased with increasing distance between the numbers. Eger et al. (2003) found that the area was activated after subjects saw the numbers even if they did not perform any digital operations. Zhang et al. (2004) studied the digital processing of Chinese subjects. The study revealed that the functions of this population were not influenced by cultural differences because the brain region was used in the same way.

Some studies have shown that the parietal lobe is not the only area that supports mathematical cognition. The results of an fMRI study by Dehaene et al. showed that digital consciousness depended mainly on the bilateral internal trenches. For example, the bilateral top trench will appear to be activated when a number and size comparison is performed. Mathematical knowledge is related to the language system and is stored in the form of words. Moreover, the researchers found that precise calculations increased the activation of the left frontal lobe and the angular gyrus region and are related to language functions. The estimated results revealed activation of the bilateral parietal cortex (Dehaene et al., 1999). Dehaene et al. (2003) found that arithmetic fact extraction in arithmetic operations activated the left temporal and parietal joint cortex. Kaufmann et al. (2008) found that children activate brain areas involving grip and finger movements, which are the left supramarginal gyrus and postcentral gyrus, respectively, during comparison tasks involving numbers. These findings suggested that children rely on counting on fingers to compare the sizes of numbers.

Based on the current research results, Dehaene et al. (2004) noted that digital processing is supported by a large network including the prefrontal cortex, parietal lobe, and temporal lobe. The bilateral parietal region, especially the area surrounding the intraparietal region, is related mainly to semantic representation. The frontal lobe, especially the left inferior frontal gyrus, has great overlap with the related brain regions responsible for verbal working memory. It is evident that there is a relationship between digital processing and language functions. The bilateral temporal lobes, especially the bilateral spindle gyrus, is related mainly to the processing of digital forms. According to the view of the multimeter sign, the above brain region is concerned with processing words, sounds, meanings and other impressions as well as different calculation tasks. For example, the estimation task is concentrated in the bilateral parietal region. The storage of mathematical facts in rehearsal is concentrated in the prefrontal area. Some digital manipulation tasks, for example, understanding complex numbers (including negative numbers and fractions), which require the processing of digital forms, are concentrated in the temporal lobe, especially with close fusiform contact (Dong et al., 2005). There are many related studies in the field of cognitive neuroscience that are not detailed here.

Additionally, Newell and Simon (1972) proposed the use of human and computer problems to solve a model called the problem-solving model. Baddeley's working memory model provides the basis for this research (Baddeley, 1986).

4.1.5 Explanation: The Theoretical Basis of Building a Cognitive Model

In conclusion, primary school children's thinking is based on specific image thinking. The lower grades of primary school prioritize material objects in the real learning concept and the basic operation process and specific image content in long-term declarative memory. The development of the internal language of children provides a theoretical basis for inspection and reflection in the process of solving problems. Cognitive neuroscience uses relevant technology to perform functional brain imaging analysis of brain activity and obtain reliable evidence about the brain mechanism of mathematical cognition activity, which has improved the scientific validity of related research. However, cognitive neuroscience and information processing analysis examine the cognitive process of solving problems at different levels. These two levels can be used to promote each other: the neuroscience data can provide a reliable basis for the cognitive model, while the cognitive model can provide a reasonable explanation for the neuroscience data.

4.2 Analysis of the Solving Process of Various Types of Primary School Mathematics Problems

The Compulsory Education Mathematics Curriculum Standards (2011 edition) divides the curriculum contents of compulsory education into four parts: "numbers and algebra," "statistics and probability," "graphs and geometry," and "synthesis and practice." In addition, the 2022 edition underscores the importance of "encouraging students to identify and formulate problems within practical contexts, and to employ a variety of strategies such as observation, conjecture, experimentation, computation, logical deduction, validation, statistical analysis, and spatial visualization to dissect and resolve these issues" (Ministry of Education of the People's Republic of China, 2022). Among the four parts, "synthesis and practice" comprehensively uses knowledge and methods from the other parts, "numbers and algebra," "graphics and geometry," and "statistics and probability," to solve problems.

According to the above classification of mathematical content, the study focuses on three main parts, namely, "statistics and probability," "figures and algebra" and "graphics and geometry," and selects typical problems for analysis. The "numbers and algebra" part concerns problems such as "numbers," "addition with different denominators" and "unary linear equations." The "figure and geometry" part concerns the problem of "mode."

4.2.1 "Numbers and Algebra" Questions

(一) "Counting" problem

"Counting" is one of the basic abilities that children must develop to learn mathematics. Many researchers have studied the "counting" problem. In research on the formation and development of primary school children's numerical concepts, Lin (1981) divided the general ability of primary school children into five grades. The first grade corresponds to the level of intuitionistic generalization and relies on the material object, teaching aids or holding one's fingers to master the concepts of numbers up to ten. In this grade, if children do not have a concrete image, the mathematical operation will be interrupted or be difficult. The development of several elementary school children's mathematical concepts follows a certain order. Regarding integers, the order of grasping the concepts of numbers within a hundred is recognition → sequence and series, composition → sapplication.

1. Analysis of the "counting" problem-solving process

"Counting" is the first concept taught in the primary school first-grade mathematics textbook (Course Textbook Institute, 2006). To solve the question of "counting" correctly, the following knowledge is required:

(1) Recognizing numbers. Recognizing numbers is the premise of "counting," and only by mastering it can one distinguish the beginning and end of the counting process. For first-grade children, their main mode of thinking is specific image thinking, so material objects are often used to teach students the methods.
(2) Number sequence. The sequence is the order relationship of numbers, such that the number after 2 is 3, the number after 3 is 4 and so on. This understanding may be simple for adults, but it is not easy for adults to teach it to children. Because adults have automated the process of counting, they need to expand the results of the automation; then, they can teach children this concept well.

Now, the problem of "count from 3 to 5" is taken as an example to analyze the problem-solving cognitive process of "counting."

(1) Understand the problem. determine that the starting point is 3 and that the end point is 5.
(2) Develop a plan. Determine the number after 3, and compare the count results with five at every turn. If the number is not 5, then continue; if it is 5, then stop.
(3) Implement the plan. Start counting from 3, activate and extract "the number after 3 is 4" in long-term declarative memory. Realize that 4 is not the end point, and keep counting. Activate and extract "the number after 4 is 5" in long-term declarative memory. Realize that 5 is the end point, after which the goal is achieved.
(4) Review. Reviewing the process can determine whether there are any problems in each link. This step can also reinforce children's awareness of numbers and sequences. The relevant links in children's long-term declarative memory can thus be connected more closely, strengthening their memory.

2. The cognitive processes of "counting."

A team led by Anderson, a professor of artificial intelligence and a psychologist at Carnegie Mellon University, studied the problem "count from 2 to 4" and wrote a corresponding ACT-R program (shown in Table 4.1). The cognitive process analysis of the problem "count" in this program is basically the same as that in this book.

The above table shows the general cognitive process of analyzing a "counting" problem, which is based on first-grade children's thinking characteristics, knowledge base, and other factors. In terms of grade level, the numeral generalization level of elementary school children is a spiraling process and an "internalized" thinking process. It proceeds as follows: in the first grade (7–8 years old), it is developed on the basis of preschool thinking and entails the generalization of specific images; in the second and third grades (8–10 years old), it transitions from the generalization of specific images to abstract images; and in the fourth and fifth grades (10–12 years old), most children are beginning to demonstrate preliminary essential abstraction generalization (Zhu, 2009).

(二) "linear equation with one unknown" problem.

A linear equation with one unknown is a typical algebraic problem, which appears for the first time in the "simple equation" unit of fifth grade mathematics teaching materials (People's Education Press). Fifth-grade children have mastered basic arithmetic operations and have previous knowledge of solving linear equations with one unknown problem.

The process of solving the equation $Ax + B = C$ can be described as follows:

(1) Understand the problem. This question involves solving an equation. The goal is to determine the numerical value of the variable x.
(2) Develop a plan. Put the expression Ax with the variable x on one side of the equation, move the other constants to the other side, and then solve for the value of x.
(3) Implement the plan. The process of solving equations can be divided into the following steps:

   ① Identify the equation: $Ax + B$ is on the left, and C is on the right.
   ② Place the expression Ax with the variable x on one side of the equation, and move the other constants to the other side of the equation. The original equation has a variant of $Ax = C\text{-}B$.
   ③ Extract D, the value of C-B, from long-term declarative memory.
   ④ Note that a variation of the original equation is $Ax = D$, and the numerical value of x is D/A.
   ⑤ Extract E, which is the numerical value of D/A from long-term declarative memory.
   ⑥ Determine the value of x to be E.

(4) Review. First, reviewing the process can help to determine whether there are any problems in each link. In addition, this step deepens the knowledge of basic

Table 4.1. Anderson "count from 2 to 4" cognitive process

ACT-R Cognitive Process	Problem-solving process analysis
(add-dm (b ISA count-order first 1 second 2) (c ISA count-order first 2 second 3) (d ISA count-order first 3 second 4) (e ISA count-order first 4 second 5) (f ISA count-order first 5 second 6) (first-goal ISA count-from start 2 end 4))	Preknowledge that is needed for problem solving Content in long-term declarative memory The correct number sequence of 1 to 5 Determine the problem goal and identify that the starting point is 2 and the ending point is 4.
(p start = goal > ISA count-from start = num1 count nil = = > = goal > count = num1	Count from 2.

(continued)

Table 4.1. (continued)

+ retrieval > ISA count-order first = num1)	Extract “the number next to 2 is 3.”
(P increment = goal > ISA count-from count = num1 - end =num1 = retrieval > ISA count-order first = num1 second =num2 = = > = goal > count = num2 + retrieval > ISA count-order first = num2 ! output! (= num1))	Extract “the number next to x is y” from long-time declarative memory and compare it with 4. If it is not 4, continue counting.
(P stop = goal > ISA count-from count = num end =num = = > -goal > ! output ! (= num))	The conditions for the settlement of the problem: if the output value is the same as the target value 4, end the process.
(goal-focus first-goal)	The goal is achieved, and the problem is solved.

arithmetic results in long-term declarative memory and further consolidates the steps of solving equations in long-term procedural memory.

The above steps describe the process that children follow the first time they solve the equation $Ax + B = C$. After a period of practice, their problem-solving process and steps are gradually automated and continuously simplified.

Anderson studied the solution of a two-step equation and compared the solutions of the Equation $7 \times x + 3 = 38$ on Day 1 and Day 5. The results showed that on Day 5, the problem-solving process was simpler than that on Day 1; on the first day, it needed 6.1 s, and on the fifth day, it needed only 4.1 s. By comparing the changes in the equation process between the first day and the fifth day, we can see that the step-by-step solution to the problem in the learning process is gradually automated (Anderson, 2005).

4.2.2 "Graphics and Geometry" Questions

Determining the "cylindrical flank area" is a typical area of knowledge included under "graphics and geometry." It corresponds to the content of the "cylinder" unit in sixth grade mathematics teaching material (People's Education Press).

Although the abstract logical thinking of sixth-grade children has reached only a certain level of development, it still has great specificity. For example, the problem of "the flank area of the cylinder" is easier to teach through the method of object teaching.

To solve this problem, the prerequisite knowledge that students need includes the following:

(1) the concept of a flank;
(2) rectangular area formula:
(3) circumference formula.

The "cylindrical flank area" problem-solving process can be described as follows:

(1) Understand the problem. Determine which part of the area needs to be calculated. The prerequisite is the ability to understand the concept of the flank, that is, to activate the related content of "flank" in long-term declarative memory.
(2) Develop a plan. Transform the problem of the "flank area of the cylinder" into the problem of the area of the rectangle.
(3) Implement the plan.

① A cylindrical flank can be cut into a rectangle, which transforms the problem into one of calculating the area of a rectangle.
② According to the known conditions, find the length and width of the rectangle.

   a. If the radius of the bottom surface of the cylinder is r, then the length of the rectangle is equal to the surface circumference of the bottom of the cylinder, that is, $2\pi r$.

b. If the height of the cylinder is h, then the width of the rectangle is the height of the cylinder h.

③ The area of a rectangle is length × width, that is, the bottom surface of the cylinder is length × height.
④ The area of the rectangle is the area of the cylinder flank.

(4) Review. Reviewing the process can help to determine whether is there any problems in each link. In addition, the review connects the rectangular area formula and the circumference formula in long-term memory more closely and further consolidates knowledge of the cylinder side expansion operation in long-term program memory.

The above analysis of the process of solving the problem of "the area of the cylinder flank" demonstrates that students may not be able to answer the questions correctly even if they possess the prerequisite knowledge. The key is how students think independently to find strategies for transforming the side of a cylinder into a rectangle. If the students had some similar experiences, such as paper cutting, parcels, etc., these experiences can help them think about the strategy of unfolding the cylinder. Moreover, teachers can help students develop these strategies consciously.

4.2.3 "Statistics and Probability" Questions

The "mode" is a typical topic in the "statistics and probability" section. It is covered in fifth grade PEP textbooks during the second semester. Ding Zuyin conducted an experimental study on children's processing of concept mastery. The results revealed that the concept mastery of primary school children presented stage features. Children in the lower grades of primary school often use "concrete examples" and "visual features" to master concepts. Students in the upper grades in elementary schools are gradually able to grasp concepts based on nonintuitive "important attributes," "practical functions" and "genus relations." Solving for the "mode" can train students to find nonintuitive "important attributes" from the data, and the "mode" is the number that occurs most frequently.

The "mode" is an abstract concept. Although fifth-grade children have some abstract thinking skills, they still need specific materials to help them understand abstract concepts. Moreover, psychological research shows that the process by which children master the conceptual system is also the process by which children apply the richness of past conceptual material to assimilate (or comprehend) profound and systematic knowledge (Zhu, 2009).

The process of solving the problem of "mode" in relation to data $\{a_1, a_2, a_3, \ldots, a_n\}$ can be described as follows:

(1) Understand the problem. Find the "mode" of the given data.
(2) Develop a plan. Find the most frequently occurring number in the data $\{a_1, a_2, a_3, \ldots, a_n\}$.

(3) Implement a plan.

① Activate the operation of counting in long-term program memory, count the number of times that a_1, a_2, a_3, …, a_n appear in the data {a_1, a_2, a_3, …, a_n}.
② Determine that the number of times that a_1, a_2, a_3, ..., a_n appear, as M_1, M_2, M_3, ..., M_n, respectively.
③ Activate the comparison operation in long-term procedural memory and compare the sizes of M_1, M_2, M_3, …, M_n to determine the maximum M_{max}.
④ M_{max} corresponds to the number a_i, and a_i is the mode of the data {a_1, a_2, a_3, …, a_n}.

(4) Review and check the solution. Reviewing the process can determine whether there are any problems in each link. It can also further consolidate the concept of "mode" in long-term declarative memory and the operations involving in calculating the "mode," the count and comparison ability in long-term procedural memory.

By analyzing the process of solving the "mode" problem, we find that the key is to determine the problem-solving strategy, that is, "the number that occurs most frequently in the data." Then, operations such as "counting," "compare," and "correspond" are all performed according to previously acquired knowledge.

Children's acquisition of concepts is a gradual process that shifts from concrete to abstract. When children begin to grasp a concept, many concepts are often isolated and have not been added to a certain conceptual system due to a lack of empirical knowledge and experience. Only by grasping a concept in the conceptual system can it be mastered better. For example, the concept of "mode" can be better mastered and its connection with other concepts, established, only if children have mastered the concepts of median and average.

4.2.4 Implications: Building an Instance Foundation for Cognitive Models

According to 2011 and 2022 editions of *The Compulsory Education Mathematics Curriculum Standards*, which classify mathematics course content in the compulsory education stage, the researcher selected typical problems such as "number and algebra," "graphics and geometry" and "statistics and probability." In accordance with the thinking features of primary school children, this book analyzes the problem-solving process for different types of mathematical problems and lays the foundation for constructing a cognitive model for solving mathematical problems in primary schools.

4.3 Cognitive Model for Solving Primary School Mathematical Problems

4.3.1 Cognitive Model

(一) The process of building a cognitive model

To build a cognitive model for solving mathematical problems in primary school, we consider the following points:

(1) Inheriting the four stages of Polya's mathematical problem-solving process: understanding the subject, developing a plan, implementing a plan, reviewing the problem-solving process, and refining each stage.
(2) Considering the basic characteristics of primary school children's thinking. These children's main form of thinking gradually transitions from concrete image-based thinking to abstract logical thinking. Even though abstract logical thinking is still, to a large extent, directly related to sensual experience, the specific image is an important component. Piaget also suggested that 7–12-year-old children's thinking occurs in the specific computing stage. Throughout primary school, intuitive teaching is an important method for capturing children's attention. The step of "finger counting" in the process of solving the "counting" problem highlights the role of "object perception" in primary school children's problem-solving process.
(3) Considering the characteristics of primary school children's memory. Primary school children have a strong ability to remember specific, distinctive features. It is based on a specific image of memory. The physical display of the "cylindrical flank area" in the problem-solving process shows that primary school children's memory is based on a "concrete object." In addition, children's working memory develops rapidly in primary school and increases with increasing grade level. After the second grade, the pace of development basically slows down (Li et al., 2003).
(4) Considering the development of primary school children's internal language. The internal language of primary school children develops gradually in school. Vygotsky's research on internal speech provides theoretical and experimental evidence for examination and reflection in the problem-solving process. Currently, the use of brain imaging in the field of cognitive psychology also confirms the existence of internal circuits in the brain. Internal language is considered a critical path for thinking (Torey, 2009). Wilson et al. used functional magnetic resonance imaging (fMRI) data from cognitive neuroscience to demonstrate the inner speech loop activation region (Wilson et al., 2004). In the process of solving this problem, the review is actually supported by internal language.
(5) Emphasizing the consolidation of memory or knowledge in problem solving. The content of working memory is consolidated over a period of time into long memory (Glickman, 1961; McGaugh, 1966). The content of the active state of

the new, reactivated memory is consolidated into a stable, inactive state after a period of time (Naderl & Hardt, 2009).

(6) The problem solution starts from the goal and ends with the goal. Moreover, it emphasizes the role of the problem situation in problem solving.

(二) Cognitive model introduction

Based on the above analysis, a cognitive model for solving mathematical problems in primary schools is constructed, as shown in Fig. 4.1.

1. Information process

Problem solving can be seen as a process, and the flow of information is described below.

(1) From object perception to short-term memory

Learners see or hear a problem to accept stimulation through the perception of the object code as nerve information. This object perception component must become the object of attention for a longer period of time, after which the object of attention enters short-term memory.

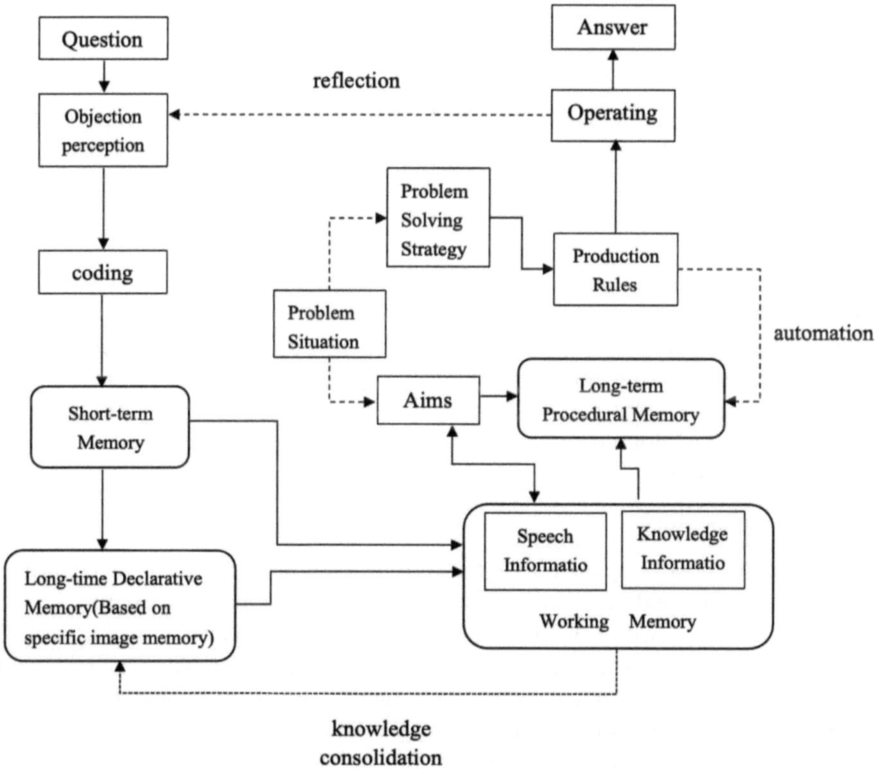

Fig. 4.1 Cognitive model of primary school mathematical problem solving

(2) From short-term memory to working memory

Short-term memory capacity is limited for adults, with an average of 7 ± 2 items (Miller, 1956). The short-term memory capacity of primary school children is lower than that of adults. It undergoes rapid development over time, becoming stable in the second grade of high school. A new object will be placed directly in working memory.

(3) From short-term memory to long-term declarative memory

If the perception is not a new object, it activates the related contents of long-term declarative memory in working memory.

(4) Long-term declarative memory

Long-term declarative memory is knowledge that the students have learned earlier. It is stored as meaningful propositions (Anderson & Bower, 1973), or complex forms of coding involving conceptual hierarchies (Quillian, 1968). The coded material in long-term memory is semantic, that is, organized according to meaning. The content stored in long, declarative memory is permanent and is retained over time (Adams, 1967). Sometimes, the extraction of information is hampered by interference between old and new content. Specific image memory is the primary form of long-term declarative memory for primary school children.

(5) Long-term procedural memory

Long-term procedural memory is a series of rules previously learned by students to produce rules for the form of storage. It includes simple rules (such as "number" rules) and complex rules (such as "dollar equation" problem-solving rules).

(6) Extraction

Extraction includes long-term declarative memory extraction and long-term procedural memory extraction. The extraction process requires clues. In the reading process, familiar words activate long-term declarative memory-related objects. Related objects are extracted to working memory. Specifically, during the problem-solving process, it will extract rules from long-term procedural memory. For example, for the two-digit addition operation, rules such as one-digit addition rules and carry rules will be extracted.

(7) Working memory

The content of working memory is the current active object, existing in the form of verbal information and image information. It contains the learner's existing knowledge and experience with new materials. Working memory may combine a learner's existing knowledge with new material to learn. For example, in explaining the concept of "the plural," "the number that occurs most often" is already in the student's working memory, but the name of the number is not yet known. The teacher then describes the concept of "the plural" to students. At this point, "the plural" and "the number that occurs most often" are combined to produce new rules, and learning takes place.

(8) Working memory of the target

The information about the problem in working memory is related to the existing schema used to represent the problem. The goal of problem representation is to understand the information contained in the problem and to determine the target.

(9) From the target to long-term procedural memory

Goals guide the problem-solving process. The target determines the goals and will activate long-term procedural memory during production.

(10) From solution to production rules

There may be more than one generative rule activated in the problem-solving process, so one of the rules is chosen to be executed under the guidance of the problem-solving strategy.

(11) Problem situation

Problem situations help students identify problem goals and choose problem-solving strategies. Similar problem situations can help learners recall special rules from previous learning and find a suitable rule for the present situation. For example, in introducing the concept of "modalities," the introduction of familiar situations such as "birthday" and the special role of "head teacher" helped students recall the rule of "taking the birthday of the largest number of months." When faced with new problem situations, students need to perform a more complex and extensive search process than when solving similar problem situations. New problem situations require a transfer of learning.

(12) From production rules to operations

The execution of production rules and production of activity patterns can be observed externally, such as writing problem-solving processes on paper or explaining ways to solve problems.

(13) Reflection

This situation occurs when the problem is solved; as the problem-solving process intensifies, it will be constantly revised. Even after the problem is complete, the process of solving the problem will be reviewed. These aspects are the external manifestations of the reflection.

(14) Knowledge consolidation

The higher the number of activated objects is, the greater the likelihood of being consolidated or strengthened is. The knowledge consolidation process reflects Hebb's law: "Activated and linked at the same time. Moreover, the more activated, the stronger the link." The result of knowledge consolidation enhances the connection of relevant objects in long-term memory, making learning useful over the long term.

(15) Automation

After completing the problem-solving process, students learn the new "chunks" formed by the rules of prior production. This "chunk" can solve new problems. For example, when students first learn different denominators, the following production rules are activated: ① different denominators ⟶add pass points, ② pass points for the least common multiple, ③ find the least common multiple, two for each prime ⟶multiply two numbers. After a period of study, the above three production rules will be combined into a new production rule: with different denominators, the denominators for the prime number and the least common multiple are calculated by taking the product of two numbers. This process is automated. The result is the emergence of "advanced rules" and other problems that can solve similar types.

(16) Summary of information flow

Figure 4.2 illustrates the structure of the cognitive problem-solving model vertically, and the right-hand column shows the machine process associated with each structure. Learners see or hear the problem to accept stimulation through the perception of the object code as nerve information. This object perception component must become the object of attention for a long time, after which it enters short-term memory. If it is a new object, it is placed directly into working memory. If the perceived object is not a new object, then the relevant content in long-term declarative memory is activated into working memory. The content of working memory comprises objects that are currently activated. It exists in the form of speech information and image information, including the knowledge and experience of learners and new materials for learning. Working memory may combine a learner's existing knowledge with new material to learn. The information about the problem in working memory is related to the existing schema used to represent the problem. The goal of problem representation is to understand the information contained in the problem and determine the target. Goals guide the process of problem solving, and the production of long-term procedural memory is activated after the goal is established. More than one production rule may be activated, and one of the rules is chosen to be executed according to the solution strategy. Production rules are implemented to generate patterns of activity that can be observed externally, for example, writing down the problem-solving process on paper or explaining a solution to the problem. Problem solving is such a situation; with the deepening of the problem-solving process, it continues to be modified and corrected. Even after the problem is solved, the entire problem-solving process is reviewed. All of these actions are external manifestations of reflection. When a new problem is solved, a high-level rule is produced.

4.3.2 *Modular Representation of the Cognitive Model*

To intuitively describe the cognitive process of problem solving, CMMPS can be simplified into the following six modules:

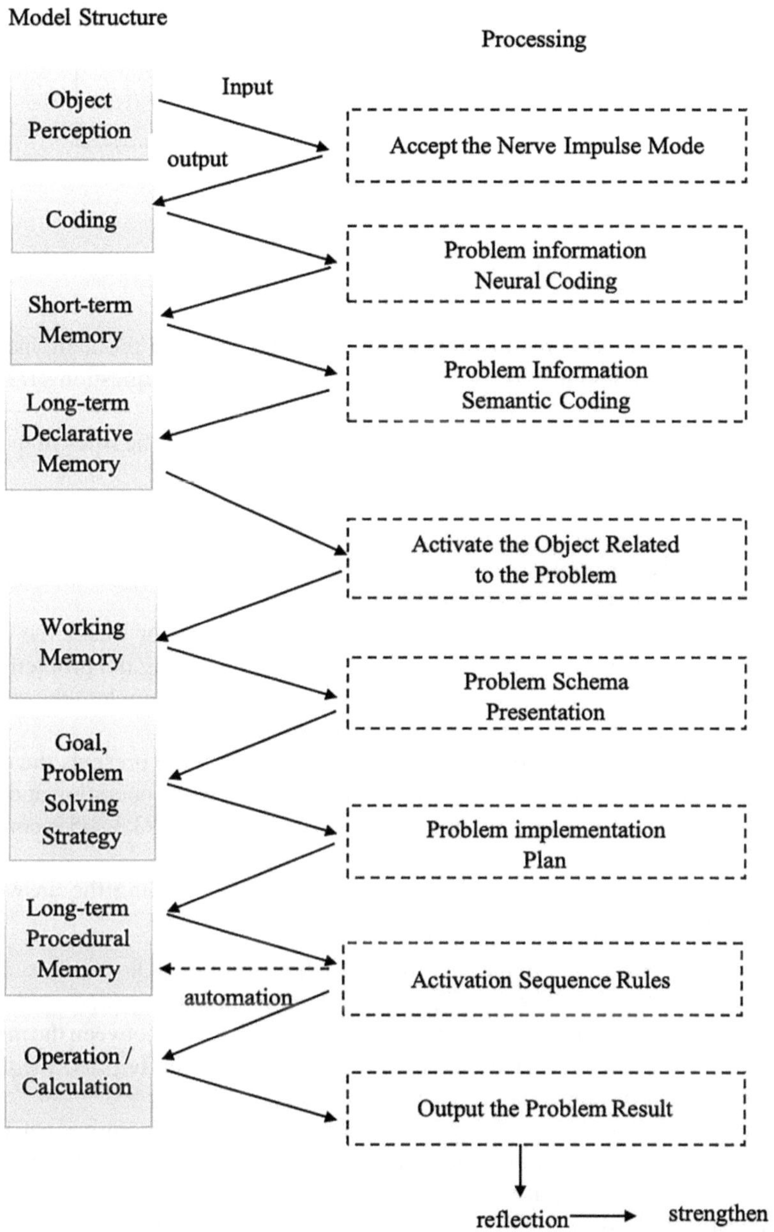

Fig. 4.2 Problem-solving process based on cognitive model

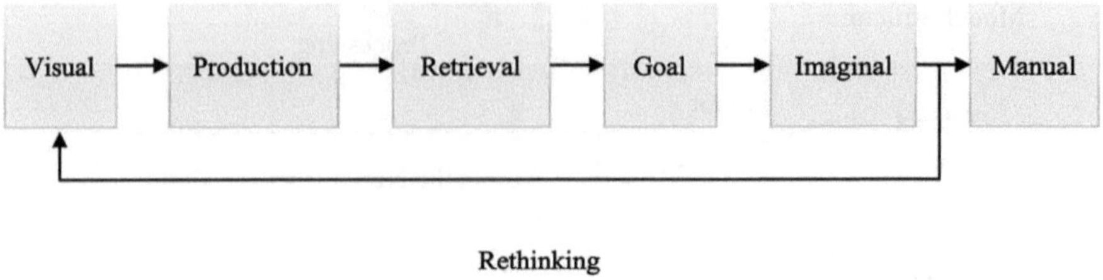

Fig. 4.3 Modular representation of CMMPS

(1) Visual module: This module retains the presentation of the problem, including the object perception and coding, such as the presentation question "1/3 + 2/5 = ?".
(2) Production module: The problem representation activates the rules in memory, including short-term memory, and the production rule.
(3) Retrieval module: This module extracts relevant information from long-term memory, including long-term declarative memory and long-term program memory. An example is the long-term declarative memory of the following facts: $5 + 6 = 11$, $1 \times 5 = 5$, and $2 \times 3 = 6$.
(4) Goal module (or control module): This module records or tracks the current purpose or intent of the problem-solving process, including the problem situation, goals, and problem-solving part of the strategy. An example is the common denominator summation problem in the strategy.
(5) Problem state module (or imaginal module): This module presents the current psychological representation of the problem, including the operation and calculation parts. For example, the problem of the original state 1/3 + 2/5 is converted to 5/15 + 6/15.
(6) Manual module: This module outputs the results, including the answer, for example, 1/3 + 2/5 = 11/15.

CMMPS is represented in modular form, as shown in Fig. 4.3.

The six modules are listed in Fig. 4.3. The problem-solving process does not involve all the modules in sequence, and the flow of information between the modules is nonlinear. The contents of the module are stored in the buffer, and the current contents of the buffer make up the working memory.

The cognitive models can be expressed in the form of $N \times 6$ cognitive matrices, as shown in Table 4.2. In the table, the numbers on the left represent the line numbers; each line represents a cognitive logic step, not an actual execution step; and the last line indicates the end of cognition. Each column corresponds to the six modules in Fig. 4.3, and the contents of each column represent the contents of a module during problem solving.

3. Description of the various stages of problem solving

Table 4.2 Cognitive matrix

	Visual	Production	Retrieval	Goal	Imaginal	Manual
1						
2						
N						

The process of solving mathematical problems in primary school is divided into four stages: understanding the problem, developing a plan, executing the plan and reviewing. The cognitive process of each stage is analyzed as follows:

(1) Understanding the problem. After students see, perceive and code the problem, they activate their long-term declarative memory of knowledge. Children's thinking in primary school is based mainly on specific image thinking. Long-term declarative memory is based on concrete objects, especially for children in the lower grades. According to the situation and existing knowledge, the brain forms a certain schema to understand the problem. The result of understanding the problem is to determine what the unknown is in the problem. What are the known data? What are the conditions? What is the goal?
(2) Developing a plan. The process from understanding the topic to formulating a plan is complex and arduous. Long-term declarative memory is activated with reference to the target to find the relationship between the known and the unknown, recall similar problems solved before, and finally obtain a solution to the problem. If there is little knowledge of the problem in long-term declarative memory, it is difficult to produce a good idea. Without knowledge, it is impossible to generate ideas. Good ideas come from children's experience and previously acquired knowledge.
(3) Executing the plan. According to the proposed solution, the objects in working memory activate the production rules in long-term procedural memory. There may be multiple activation rules, but only one production rule can be executed at the same time. The result of the production rule execution is the operation or calculation, which constitutes the answer to the problem.
(4) Reviewing. This step helps consolidate content in working memory into long-term declarative memory. Multiple production rules are automatically generated as production rules and consolidated into long-term procedural memory. This step involves reflecting on the process of solving the problem, checking whether the result is correct, understanding the approach to solving the problem, and cultivating the ability to solve the problem.

4.3.3 Cognitive Model Characteristics

CMMPS is based on the thinking characteristics of primary school children, taking into account the rules of primary mathematics. It has the following features:

(1) Highlights the importance of the problem situation.

Pupils have less abstract knowledge. It is important to understand the problem situation to solve the problem. Problem situations can help students understand problems and translate applied problems into computational problems. The calculation of the problem is relatively simple for students. Moreover, it is easy to correct the answer. The problem situation should be related to the actual life of the student.

(2) The content of long-term declarative memory is limited, and knowledge-based and concrete-based knowledge is dominant. As students progress through grade levels their abstract knowledge increases gradually.

(3) There is very little content regarding strategies and steps for solving problems in junior middle school in long-term procedural memory. However, it increases with problem solving.

(4) In terms of the production rule set, the math problems in the lower grades of primary school involve mainly simple production rules. In later grades and with greater knowledge, some simple production rules form "chunks," producing a new production rule that is preserved in long-term procedural memory.

(5) Cognitive process refinement for problem solving can be used for both diagnostics and automation.

4.3.4 A Few Notes on the Cognitive Model

(1) The cognitive model describes the problem-solving process in terms of thinking and memory.

Although the model describes the thinking process of problem solving, it is constructed at the cognitive level. This design allows for a more detailed explanation of the thinking process from the memory level. This study provides more specific and operational methodological guidance for teaching.

(2) Problem solving is a nonlinear process.

Events such as the following may occur during the problem-solving process: students may come up with a very good solution and skip all the preparatory steps to obtain the answer to the question directly without going through the stages of the cognitive model. However, if a student ignores a certain phase of problem resolution and does not generate a good idea, it will be difficult to answer the question correctly. If students do not understand the problem and begin to solve it, they will not answer the question correctly. In the process of implementing the plan, if students check each step, they can avoid many mistakes. If students do not re-examine or consider the solution again, some of the best results may be lost. Problem solving may involve different processes depending on students' knowledge and the problems.

(3) The cognitive model does not consider the student's "will," "willpower" or other emotional factors in the process of solving problems.

In the process of solving a problem, it is insufficient to understand only the subject. Students need to have the will to solve the problem. If students do not have a strong desire to solve problems, they may give up when they encounter difficulties in the process of solving problems. Moreover, it will be impossible to solve a difficult problem. Only with such a desire will it be possible to answer the question correctly. However, emotional factors in the problem-solving process are very complex and are not the focus of this study.

4.3.5 *Educational Significance of the Cognitive Model*

The cognitive model of primary school mathematical problem solving has a great influence on the design and diagnosis of problems in the teaching process.

(1) The model shows that the problem solution consists of several stages. Moreover, each stage contains several internal processing processes. To produce a certain learning result, the design process should be based on internal processing. For example, design issues, the problem context and student life are linked according to the cognitive characteristics of primary school students.
(2) Diagnose issues that arise during problem solving and provide interventions to ensure that learning occurs. The result of problem solving cannot be judged by a simple "right" or "wrong." The cognitive model is used to analyze the internal process that leads to problem solving and to propose questions that stimulate the memory of the relevant rules to guide students to answer the questions correctly. For example, in the study of the "plural" concept, "birthday" situations and "teacher in class" roles are provided to guide students in determining "the month with the largest number of birthdays" rule.
(3) Explain the problem-solving behavior and expecting learning results. The cognitive model can analyze the internal processing involved in generating problem solutions. This model infers the activation of long-term declarative and procedural memories based on internal processes, explains problem-solving behavior, and anticipates learning outcomes.

4.4 Summary

The 2011 and 2022 editions of *The Compulsory Education Mathematics Curriculum Standards* note that the design of a math curriculum in compulsory education should fully consider the characteristics of students' math learning at this stage and meet students' cognitive and psychological characteristics. To implement the concept and intention of curriculum standards in teaching practice, this chapter focuses on the construction of a cognitive model of mathematical problem solving in primary schools (Ministry of Education of the People's Republic of China, 2012a, 2012b, 2022).

The cognitive model is the basis of problem-solving cognitive process analysis. This approach considers not only the cognitive rules and characteristics of primary school students but also the mathematics curriculum content characteristics of the primary school. This chapter first analyzes the psychological characteristics of primary school children, including the characteristics of the development of thinking, the development of internal language features, memory development characteristics, and cognitive neuroscience in their mathematical cognition. This analysis lays the theoretical foundation for constructing the cognitive model.

On this basis, we analyze the typical problem-solving processes in number and algebra (such as the number problem and one dollar equation problem), graphics and geometry (such as the cylindrical side area problem), and statistics and probability (such as the mode problem) problems. We further refine Polya's mathematical problem-solving model and construct a cognitive model of elementary mathematical problem solving. We subsequently analyze the characteristics of the cognitive models, define the scope of application of cognitive models and discuss the value and significance of cognitive models for mathematical classroom teaching.

A cognitive model of primary mathematical problem solving provides a scientific basis and reference for achieving cognitive problem-solving simulation and discussing the application of cognitive analysis and simulation in math classroom teaching.

References

Adams, J. A. (1967). *Human memory* (pp. 79). McGraw-Hill.

Anderson, J. R., & Bower, G. H. (1973). *Human associative memory* (pp. 102). V. H. Winston.

Anderson, J. R. (2005). Human thumb evaluation in reinforced cognitive artificial. *Cognitive Science, 29*, 313–341.

Baddeley, A. D. (1986). *Working memory* (pp. 17). Oxford University Press.

Baddeley, A. D., & Logie, R. H. (1999). Wording memory: The multiple-component model. In A. Miyake & P. Shah (Eds.), *Models of working memory: Mechanisms of active maintenance and executive control* (pp. 28–61). Cambridge University Press.

Chen, G. P., & Wang, X. L. (2005). Transient study of short-term memory and strategy development. *Psychological Science*, 4.

Course Textbook Institute. (2006). *Compulsory education curriculum standard textbook fifth grade mathematics* (pp. 2). People's Education Press.

Dehaene, S., Molko, N., Cohen, L., & Wilson, A. (2004). Arithmetic and the brain. *Current Opinion in Neurobiology, 14*, 218–224.

Dehaene, S., Piazza, M., Pinel, P., & Cohen, L. (2003). Three parietal circuits for number processing. *Cognitive Neuropsychology, 20*, 487–506.

Dehaene, S., Spelke, E., Pinel, P., Stanescu, R., & Tsivkin, S. (1999). Source of mathematical thinking: Behavioral and brain-imaging evidence. *Science, 284*, 970–974.

Dong, Q., Zhang, H. C., & Zhou, X. L. (2005). Mathematical cognition: The research results of brain and cognitive science and its educational enlightenment. *Journal of Beijing Normal University (Social Science Edition)*, 3.

Eger, E., Sterzer, P., Russ, M. O., Giraud, A.-L., & Kleinschmidt, A. (2003). A supramodal number representation in human intraparietal cortex. *Neuron, 37*, 719–726.

Gazzaniga, M. S. (1998). *Cognitive neuroscience* (Z. Shen, Trans.). Shanghai Education Press.

Glickman, S. (1961). Perseverative neural processes and consolidation of the memory trace. *Psychological Bulletin, 58*, 218–233.
Kaufmann, L., Vogel, S. E., Wood, G., Kremser, C., Schocke, M., Zimmerhackl, L.-B., & Koten, J. W. (2008). A developmental fMRI study of nonsymbolic numerical and spatial processing. *Cortex, 44*, 376–385.
Li, D. M., Liu, C., & Li, G. Y. (2003). Digital working memory breadth of life-long development and its role factors. *Acta Psychologica Sinica.*
Lin, C. D. (1981). The study of the development of primary school children's concepts and computing capabilities. *Psychological Journal*, 3.
McGaugh, J. L. (1966). Time-dependent processes in memory storage. *Science, 153*, 1351–1358.
Miller, G. A. (1956). The magical number seven, plus or minus two: Some limits on our capacity for processing information. *Psychological Review, 63*, 81–97.
Ministry of Education of the People's Republic of China. (2012a). *Curriculum standards of compulsory education* (2011 ed.). Beijing Normal University Press.
Ministry of Education of the People's Republic of China. (2012b). *Compulsory education mathematics curriculum standards* (2011 edition). Beijing Normal University Press.
Ministry of Education of the People's Republic of China. (2022). *The compulsory education mathematics curriculum standards* (2022 ed.). Beijing Normal University Press.
Ministry of Science and Technology of the People's Republic of China. (2006). *National medium and long term plan for science and technology development (2006–2020)*. Retrieved from http://www.most.gov.cn/mostinfo/xinxifenlei/gjkjgh/200811/t20081129-65774.htm
Nader, K., & Hardt, O. (2009). A single standard for memory: The case for reconsolidation. *Nature Reviews Neuroscience, 10*, 224–234.
Newell, A., & Simon, H. A. (1972). *Human problem solving* (pp. 19). Prentice Hall.
Pinel, P., Dehaene, S., Rivière, D., & Le Bihan, D. (2001). Modulation of parietal activation by semantic distance in a number comparison task. *NeuroImage, 14*, 1013–1026.
Poldrack, R. A. (2008). The role of fMRI in cognitive neuroscience: Where do we stand? *Current Opinion in Neurobiology, 18*, 223–227.
Qian, H. F., Zhang, L. X., & Li, S. C. (1989). Preliminary study on the development of short-term memory in primary school children. *Psychology*, 1.
Qin, Y., Anderson, J. R., Silk, E., Stenger, V. A., & Carter, C. S. (2004). Change in brain activation patterns during children's practice in algebra equation solving. *Proceedings of the National Academy of Sciences of the United States of America, 101*, 5686–5691.
Quillian, M. R. (1968). Semantic memory. In M. Minsky (Ed.), *Semantic information processing* (pp. 19). MIT Press.
Torey, Z. (2009). *The crucible of consciousness: An integrated theory of mind and brain* (pp. 112). MIT Press.
Vygotsky, L. S. (2010). *Thinking and language* (W. Li, Trans.). Peking University Press.
Wilson, S. M., et al. (2004). Listening to speech activates motor areas involved in speech production. *Nature Neuroscience, 7*, 701–702.
Zhang, H. C., Dong, Q., Jin, Z., & others. (2004). Multiplication by mental number line: An fMRI study. *The 28th International Congress of Psychology.*
Zhou, X. L., Chen, C. S., et al. (2009). Event-related potentials for simple arithmetic in Arabic digits and Chinese number words: A study of the mental representation of arithmetic facts through notation and operation effects. *Brain Research, 1320*, 212–224.
Zhou, X. L., Chen, C. S., Chen, L., & Dong, Q. (2008). Holistic or compositional representation of two-digit numbers? Evidence from the distance, magnitude, and SNARC effects in a number matching task. *Cognition, 106*, 1525–1536.
Zhou, X., Chen, C., Zhang, H., Chen, C., Zhou, R., & Dong, Q. (2007). The operand-order effect in singledigit multiplication: An ERP study of Chinese adult. *Neuroscience Letters, 414*, 41–44.
Zhu, Z. X. (2009). *Children's psychology* . People's Education Press.

Chapter 5
Primary School Mathematical Problem-Solving Cognitive Simulation

5.1 Cognitive Simulation Basis

There are many successful cases in which computer simulation methods are used to study the internal process of problem solving. These cases provide a fundamental basis for the use of these methods to study the cognitive process of students' mathematical problem solving.

Newell and Simon wrote the first computer program to simulate the human problem-logic theorist (LT). It successfully mimicked the human cognitive process of proving the symbolic logic theorem (Newell & Simon, 1956). LT proved all 52 theorems in Whitehead's *Mathematical Theory,* which simulates the problem-solving process of human heuristic search.

Newell and Simon developed the General Problem Solver (GPS) program (Newell et al., 1959). The program is based on the "means–purpose analysis" method of preparation. The program successfully simulated many different types of problems, such as theorem proofs, the tower of Hanoi and missionaries, and the savages' problem of crossing the river. The GPS system contains long-term memory as a knowledge base that stores various problems related to solving knowledge and different operators. It also features short-term memory in a serial manner for a variety of information operations. A production system characterizes the internal knowledge of GPS. The program uses the search of the problem space and the "means–purpose analysis" method to reduce the current state and the target state differences And eventually reach the target state.

Gelernter et al. (1960) developed a computer program that simulated the human geometry theorem—Geometry Machine. Hiller and Isaacson (1959) developed a computer program that simulated the creation of musical compositions. Newell et al. (1958) developed a simulated human chess program. Newell et al. (1957) developed computer programs that modify many aspects of themselves based on experience and thus achieve "learning."

X. Wei, *Artificial Intelligence in Mathematics Education*, Smart Computing and Intelligence, https://doi.org/10.1007/978-981-95-1405-2_5

Simon (1986) conducted computer simulations of thinking and problem-solving behaviors, such as insight and understanding. He believes that computer simulation is a powerful tool to predict and explain a large number of thinking phenomena.

Anderson et al. (2008) used ACT-R to simulate the solving process of the algebraic equation "$7x + 3 = 38$."

Wu Wen jun, a Chinese scholar, put forward a mathematical algorithm for proving the geometry theorem, which is called the "Wu method" (Wu, 1984). Academician Zhang et al. (1955) improved on this method, creating a new algorithm to automatically solve the problem of nearly any geometric proof. However, both scholars studied automatic problem solving from the perspective of mathematics and did not consider students' cognitive process of problem solving.

The cognitive simulation of mathematical problem solving is based on a certain cognitive model to write computer programs to simulate the students' cognitive process of solving mathematical problems. This design enables the computer to achieve similar results as the students do. The internal process of student problem solving cannot be obtained directly. Computer simulation can visualize the internal process and has become a common method in this field.

5.2 Cognitive Simulation Tools

5.2.1 ACT-R Tools

Adaptive control of rational thinking (ACT-R) has been a well-known cognitive simulation tool for many years in the Cognitive Science Laboratory under the leadership of Professor Anderson, a famous cognitive psychologist at Carnegie Mellon University. It embeds ACT-R theory, and the programming language is Common Lisp. The current version is ACT-R 7, version 7.27.9. Its internal structure and parameter setting are based on a large amount of psychological experimental data. Many ACT-R data have been verified by NMR experiments. Like a programming language, ACT-R is a framework. For different tasks, researchers can combine the cognitive view of ACT-R to strengthen their assumptions concerning specific tasks and establish models with ACT-R. Assumptions can be verified by comparing the results of the model with the results of people completing the same task. ACT-R has been widely used to simulate different aspects of human cognitive behavior, such as the Hanoi tower problem, language understanding, pattern recognition, memory, and simple geometric proofs (Wei et al., 2011).

5.2.2 *ACT-R Application Field*

ACT-R has facilitated the creation of models in areas such as learning and memory, problem solving and decision-making, language and communication, perception and attention, cognitive development, and individual differences. In addition to its application in the field of cognitive psychology, ACT-R has been successfully applied in other fields, as shown in Fig. 5.1.

(1) Human–computer interaction: to produce user models that can assess different computer interfaces;
(2) Cognitive tutoring systems: to "guess" the difficulties that students may have and provide focused assistance;
(3) Neuropsychology: to interpret FMRI data.

ACT-R has been successfully applied in mathematical problem-solving simulations. By studying the ACT-R model of algebraic equations, Anderson preliminarily mapped the structural components of ACT-R theory to a corresponding brain area (Anderson, 2005). ACT-R emphasizes the measurement of behavior levels, such as keystroke time and eye movement mode. Anderson et al. used brain imaging technology to verify the relationship between the ACT-R model and brain areas in the process of solving equation tasks.

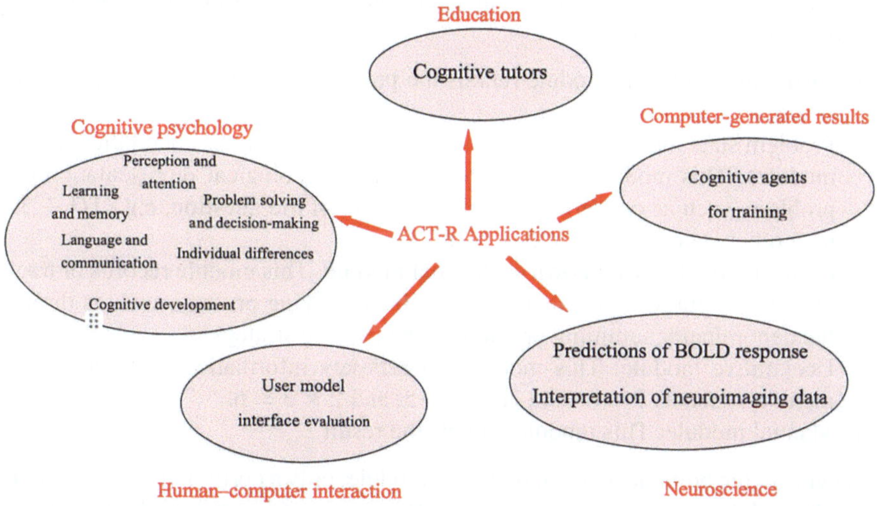

Fig. 5.1 Application field of ACT-R

5.2.3 ACT-R Cognitive Neurology Basis

ACT-R studies have a certain cognitive neurology basis. Qin et al. (2003) and Anderson et al. (2004) found that although parietal and prefrontal activities are highly correlated, the activities of these two regions are still distinct. They also confirmed that the prefrontal lobe is more strongly associated with knowledge extraction and that the parietal lobe is more strongly correlated with characterization (problem status) changes. Sohn et al. also found that the prefrontal, rather than the parietal, cortex is associated with personal knowledge extraction (Sohn et al., 2005).

Regarding procedural memory in ACT-R, Ashby and Waldron (2000) studied the neuropsychological basis of sample learning. Hikosaka et al. (1999) studied the neural network basis of learning sequence programs, and the results revealed that the basal ganglia are associated with procedural memory. In addition, D'Esposito et al. (1995) demonstrated that the anterior cingulate cortex (ACC) is consistent with the central executive system in the Baddeley working memory model (Baddeley, 1986).

5.3 Cognitive Simulation

Anderson studied the process of solving algebraic equations. Cognitive processes occur through interactions with five independent modules (Anderson, 2005). These five modules are as follows:

(1) Visual module: This module retains the problem characterization, such as $1/3 + 2/5$.
(2) Problem state module or problem state module (sometimes called the imaginal module): This module presents the current psychological representation of a problem, such as converting the original state of the question, e.g., $1/3 + 3/5$, to another state, such as $5/15 + 6/15$.
(3) Control module, also known as the goal module: This module records or tracks the current purpose or intent of the problem-solving process, such as through the denominator summing problem in the passing strategy.
(4) Declarative module: This module extracts key information from declarative memory, such as $5 + 6 = 11$, $1 \times 5 = 5$, and $2 \times 3 = 6$.
(5) Manual module: This module outputs the result.

Owing to the bottleneck of the sequence module, only a portion of the information in each module can be entered into the buffer connected to the module, such as a perceived object, a representation of the problem state, and the state of control. Each buffer has only one so-called "knowledge block" in ACT-R. The knowledge block contains question status information and control status information. Only one production rule can be activated at a time in ACT-R.

Elementary mathematics can be divided into two types of knowledge: procedural knowledge (referred to as PK) and declarative knowledge (referred to as DK).

According to the classification of knowledge, "addition with different denominators" is an example of typical procedural knowledge, and "mode" is an example of typical declarative knowledge. In this study, two points of knowledge were selected: typical problems were designed, the cognitive process of problem solving was analyzed, and a simulation was performed.

5.3.1 Cognitive Simulation of the Procedural Knowledge Problem Solving

(一) A typical topic

The content of our analysis is the knowledge point of "different denominators" in the fourth unit of the fifth grade of elementary school, "meaning and nature of the scores." The textbooks used are compulsory education curriculum standard experimental textbooks published by the People's Education Press (2nd edition, Oct. 2006). The teaching goal for the knowledge "addition with different denominators" is to learn how to perform addition with two different denominators. This problem is a typical problem involving primary school mathematics procedural knowledge.

Before learning "addition with different denominators," students already know the features of natural numbers n 2, 3, and 5 to be able to determine the common multiple and the least common multiple. With a natural number of 1 to 100, students can find all multiples of natural numbers within 10 and the common multiple and the least common multiple of two natural numbers within 10.

According to the "addition with different denominators" knowledge points and the characteristics of the students, we designed the following questions:

> Please color the rectangular paper (as shown in Fig. 5.2). One-third of the paper should be colored yellow, and 2/5 of the paper should be colored black. The same color cannot be used for another area (the yellow area cannot be colored black, and the black area cannot be colored yellow). What percentage of the paper is yellow and black?

(一) Cognitive process analysis

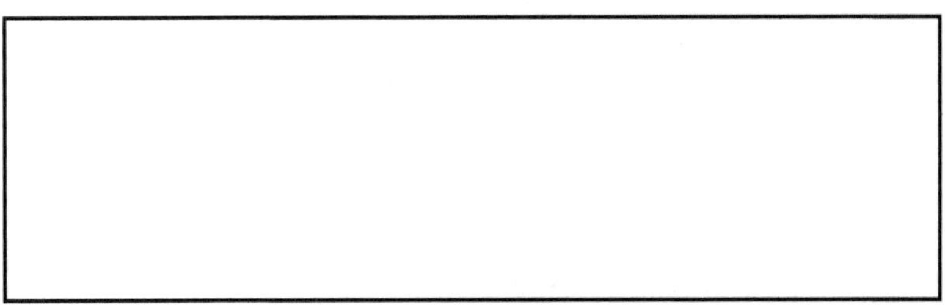

Fig. 5.2 Rectangular paper

The problem of "addition with different denominators" is analyzed by CMMPS. The cognitive process of problem solving can be described as follows:

(1) Students visualize the problem and then activate the relevant objects in long-term declarative memory. They achieve the intended understanding and identify the goal as the addition of different denominators, that is, "1/3 + 2/5 = ?", thus completing the conversion from an application problem to a calculation problem.
(2) To solve the problem "1/3 + 2/5 = ?", activate the production "addition with different denominators, find the least common multiple." The target is determined to be the least common multiple of 3 and 5.
(3) Require the least common multiple of 3 and 5 to activate the least common multiple of productions "3 and 5? ➡ 3 × 5" to extract the fact "3 × 5 = 15" in long-term declarative memory;
(4) After the least common multiple is obtained, convert the different denominators into the same denominator, that is, the pass points, and divide "1/3" and "2/5" into "5/15" and "6/15," respectively.
(5) After the pass, transform the problem of addition with different denominators into addition with the same denominator to activate the production of the rule "Add the numerators after calculating the same denominator";
(6) Extract the long-term declarative memory "5 + 6 = 11," and determine the result is "11/15." The cognitive process is complete.

To vividly represent the problem of "adding different denominators" to solve the cognitive process, the analyzed results are expressed in modular form in Table 5.1.

In Table 5.1, in addition to the abovementioned five modules of the ACT-R (retrieval corresponding declarative module), an additional production module is added. This strategy is used in problem-solving process. In this module, the content is the production rule activated in the problem-solving process. Each column represents the content of a module in the problem-solving process. The leftmost column of numbers in Table 5.1 indicates the row number, and each line represents the cognitive logical step, which is not the same as the actual solution step. The last line indicates that the cognitive process is complete; that is, the problem-solving process is complete. Each column shows the content of each module at different times.

(二) Cognitive simulation

The chunk used to encode the propositional information is the same as the knowledge block in ACT-R. The problem is understood in the form of a chunk, which contains multiple slots. The title "addition with different denominators" can be expressed as follows:

Is a mathematical problem,
The object paper has two color areas.
Known 1: yellow area value is 1/3 of the paper
Known 2: black area value is 2/5 of the paper
Goal: a sum, addend 1 yellow area and addend 2 black area

Table 5.1 Analysis of the "addition with different denominators" cognitive problem-solving process

	Visual	Production	Retrieval	Goal	Imaginal	Manual
1	Visual coding					
2			Relevant semantic knowledge in long-term declarative memory			
3				1/3 + 2/5 = ?	1/3 + 2/5 = ?	
4	Coding 1/3 + 2/5 = ?					
5		Addition ➡ with different denominat-ors Find the least common multiple				
6				Find the least common multiple		
7	Coding 3 and 5					
8		The least common multiple of 3 and 5 ➡ is 3 × 5				
9			3 × 5 = 15			
10						The least common multiple is 15
11				Reduce fractions to a common denominator		
12		1/3 ➡5/15 2/5 ➡ 6/15				
13			3 × 5 = 15 1 × 5 = 5 5 × 3 = 15 2 × 3 = 6			

(continued)

Table 5.1 (continued)

	Visual	Production	Retrieval	Goal	Imaginal	Manual
14						5/15 6/15
15					5/15 + 6/15 = ?	
16				Addition with the same denominator		
17	5/15 + 6/ 15 = ?					
18		Add with the same denominator Given ➡a common denominator, sum the numerators				
19			5 + 6 = 11			
20					5/15 + 6\15 = 11/15	
21						Sum the fractions (5/15 + 6/ 15) = 11/ 15
22						End

The first line indicates that the problem is a mathematical problem. The second line indicates that the object is paper, and there are two color areas. The third line shows the known condition 1: one-third of the paper is yellow. The fourth line shows the known condition 2: two-fifths of the paper are black. The last line indicates that the goal is to seek two sums of the numbers: addend 1 is the yellow area, and addend 2 is the black area.

The Lisp program was written (see Appendix 1) based on the analysis of the cognitive solving process of the above problem of "addition with different denominators." It was simulated in ACT-R, and the minimum time interval was 0.05 s (default). The problem-solving cognitive process simulation of this problem is shown in Fig. 5.3.

Figure 5.3 shows that setting goals in the problem-solving process is a crucial step. It begins with a definite goal. The intermediate process is the constant conversion of the problem state and finally ends with the goal. "Procedural" refers to procedural knowledge, that is, production. The extraction of procedural knowledge involves activating the production of only one rule at a time. "Declarative" refers to declarative knowledge, expressed in the form of a knowledge chunk. The extraction of declarative knowledge is the operation of the knowledge chunk.

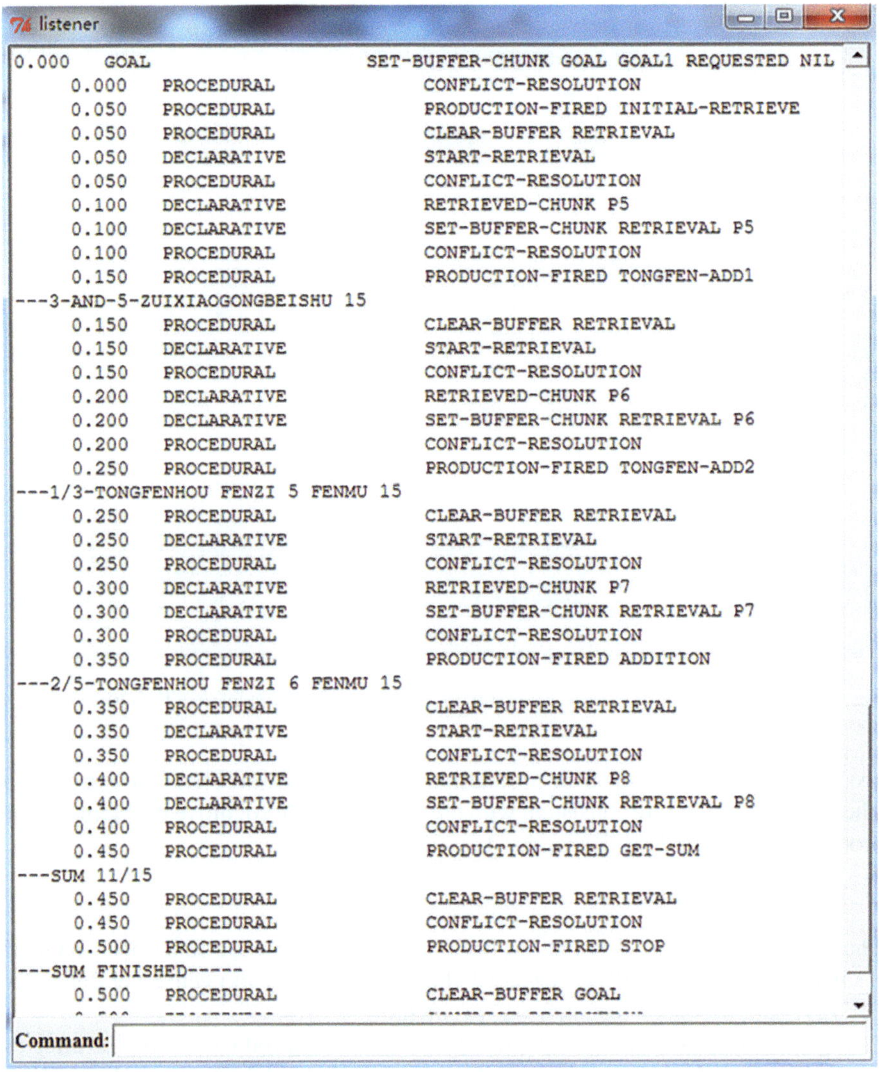

Fig. 5.3 "Addition with different denominators" cognitive problem-solving process simulation

The cognitive trace can be visualized via ACT-R. Figure 5.4 shows the cognitive trace used to solve the problem of "addition with different denominators." In Fig. 5.4, the leftmost column is the module in ACT-R. For example, there is a retrieval module, an imaginal module, a visual module, a production module, and a goal module, among others. On the right, the contents of each module are displayed according to the time sequence (the basic event unit is the default value of 0.55 s). The red area shows the contents of the extraction module, that is, the declarative knowledge extracted when solving the problem. The yellow area shows the contents of the

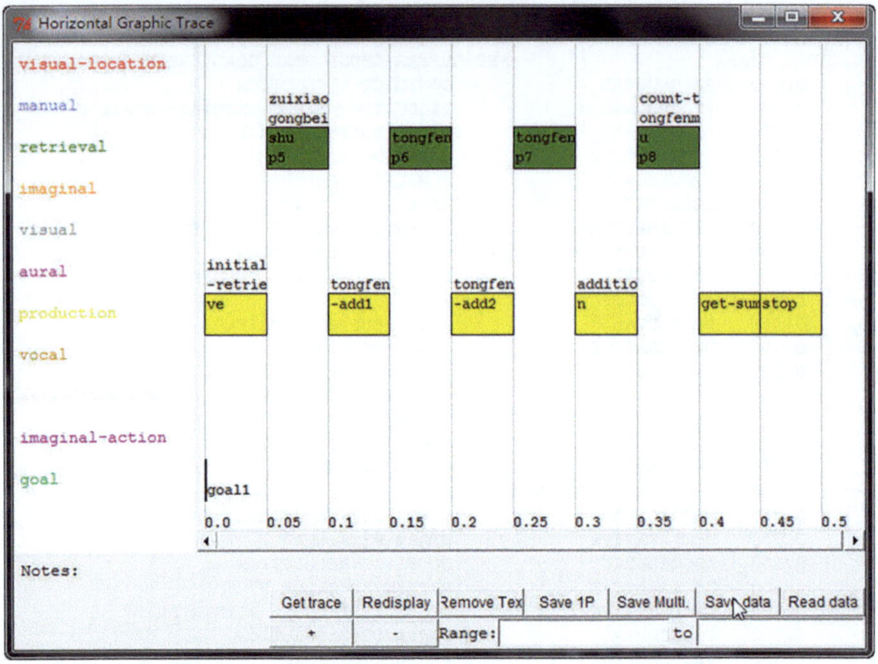

Fig. 5.4 Cognitive trace to solve the problem of "addition with different denominators"

production module, which is the production activated in the problem-solving process. These visualizations are consistent with the analysis of the cognitive problem-solving process of "1/3 + 2/5 = ?" presented in Table 5.1.

(三) Activated brain areas

The components of ACT-R map to brain regions that can use functional magnetic resonance imaging (fMRI) to record the brain's blood oxygen level-dependent (BOLD) data for the "alias addition" problem.

Figure 5.5 shows the BOLD data changes in the "production" buffer after the "addition with different denominators" of the ACT-R model. The left column shows all buffers in the problem model. One of the buffers is selected, and the blood oxygen level-dependent data in that buffer are displayed on the right. The horizontal axis represents the time, and the default time interval is 1.5 s. The vertical axis represents the variation range, and the minimum value is "0.0," which means no activation; the higher the value is, the higher the activation is, and the maximum is "1.0." The figure clearly shows that the activation of the content in the "production" buffer over time was most prominent between 5 and 6 s and then slowly decreased until 13.5 s, when it reactivated.

The brain activation areas for the "addition with different denominators" problem-solving process is shown in Fig. 5.6. The left column is marked with a different color buffer. The right image shows the brain activation area at a given time in the process.

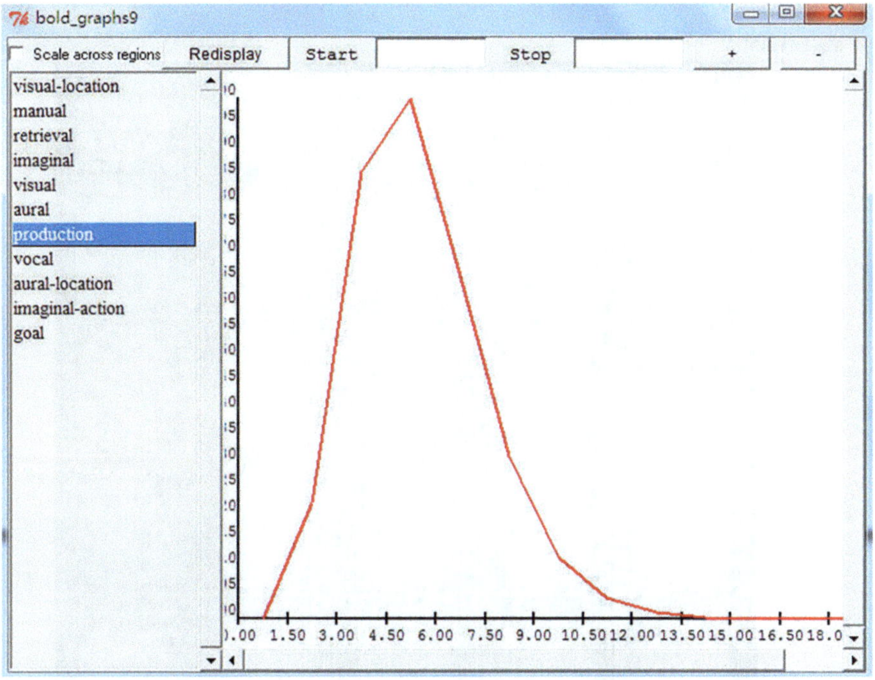

Fig. 5.5 Activation of the "production" buffer during the "addition with different denominators" problem-solving process

The color of the box corresponds to the color of the buffer on the left. The brightness of the area indicates the degree of activation. The brighter the area is, the more it is activated.

As shown in Fig. 5.6, the extraction of the contents of the image buffer is closely related to the activation of the parietal cortex. This finding is consistent with the findings of the studies by Pinel et al. (2001), Eger et al. (2003) and Zhang et al. (2004) In those studies, the parietal cortex of the subjects was significantly activated when they saw the numbers or performed digital processing. The default time for an image in ACT-R is 200 ms.

The retrieval buffer is responsible for extracting declarative memory. It is associated with the activation of the prefrontal cortex. This finding is consistent with the findings of Qin et al. (2003), Anderson et al. (2004), and Sohn et al. (2003, 2005). The prefrontal, rather than the parietal, cortex correlates with the extraction of personal knowledge. The extraction time in ACT-R is a free variable.

The procedural buffer is responsible for the extraction of procedural knowledge and is closely linked to the activation of the basal ganglia. This finding is consistent with that of Hikosaka et al. (1999).

Figure 5.7 shows the solution to the problem of "addition with different denominators" in the brain model in the form of a three-dimensional map of the brain

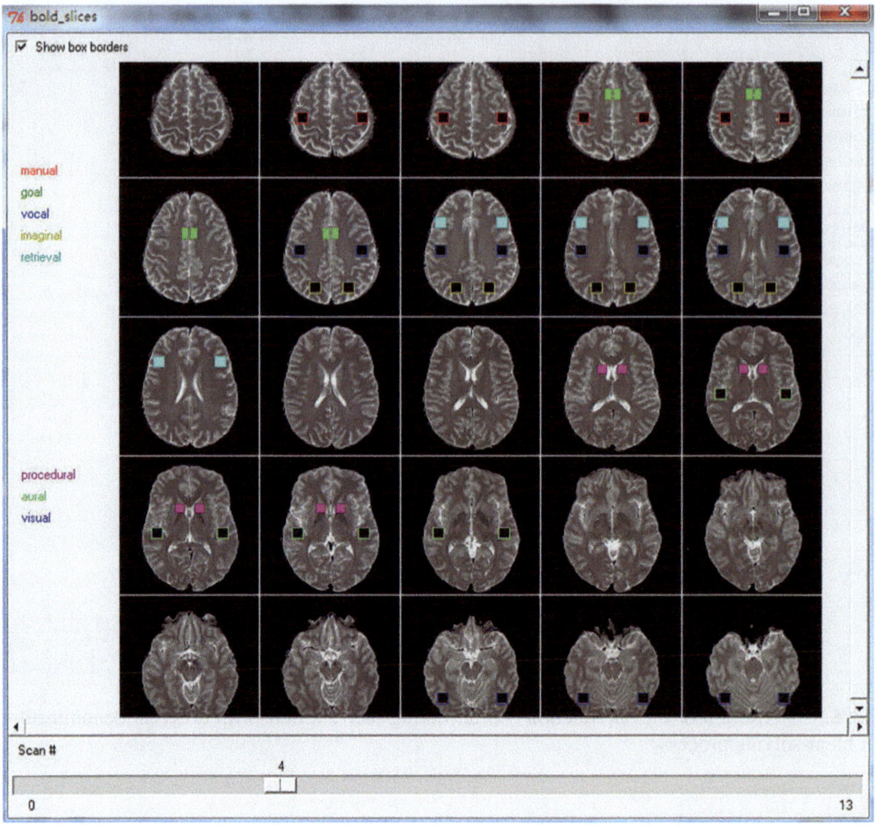

Fig. 5.6 Brain activation areas for the "addition with different denominators" problem-solving process

activation areas. In Fig. 5.7, "0. 0–1.0" represents the brightness value. A value of "0" means that it has not been activated and that the area is black; the more active it is, the closer the value is to "1," and the brighter the area is. The left side of the graph shows the buffer in a different color, and the number to the right of the buffer is the activation level. The right side of the graph is the brain activation area, shown in the same color as the left module.

As shown in Fig. 5.7, the target, extraction and production buffers are all activated to different extents in the process of solving the "addition with different denominators" problem, in which the target buffer is activated at a maximum of 0.981, which is close to the maximum value. The corresponding relationship between the buffer zone and the brain area is the same as the result shown in Fig. 5.6 and is supported by existing research, so it is not repeated here.

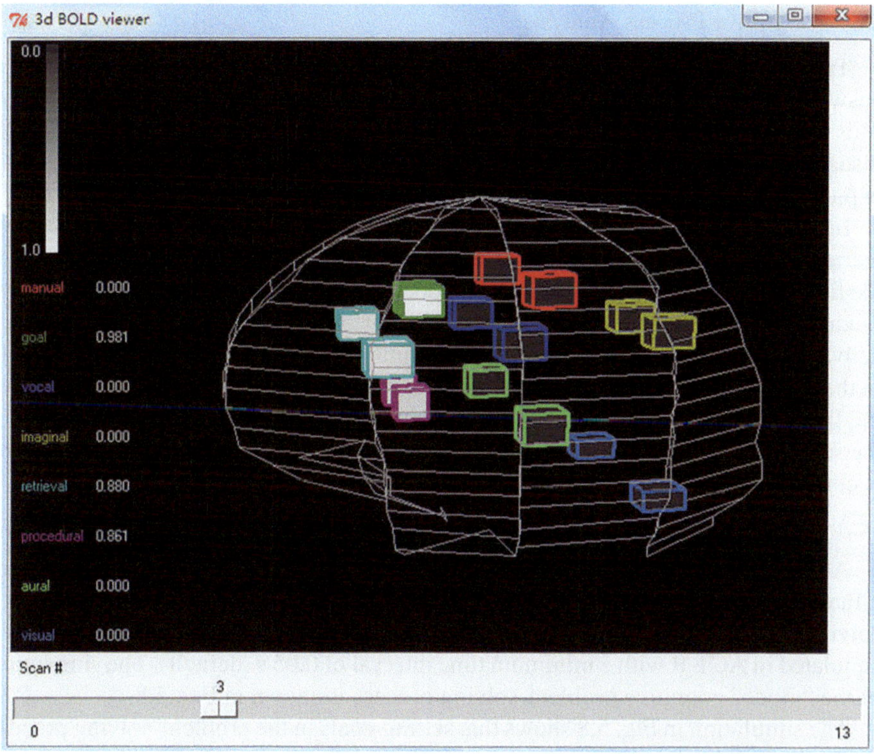

Fig. 5.7 Three-dimensional display of the brain activation areas for the "addition with different denominators" problem-solving process

5.3.2 Cognitive Problem-Solving Knowledge Simulation

(一) A typical problem

The content of our analysis is the "mode" knowledge point in "statistics," the sixth unit of the fifth grade in elementary school. The materials used for the People's Education Press compulsory education curriculum standard experimental textbooks (2nd edition, October 2006). The teaching goal of the "mode" knowledge point is to enable students to understand and master the concept of "mode," which is a typical problem that requires declarative knowledge.

According to the "mode" concept and the characteristics of the students, we designed the following problem:

"The school agreed to set aside five (six) classes next year for a birthday celebration, but only the birthdays of students born in a certain month can be celebrated. Imagine you are the class teacher:

(1) How would you choose the month?
(2) Which month do you think should be chosen?"

(二) Cognitive Process Analysis

The "mode" content is declarative knowledge. In the process of seeking the answer, we need statistical knowledge, numerical procedures and other knowledge. This knowledge can be extracted from students' long-term procedural memory. To visualize the cognitive process of solving for the "mode," the analytical process is expressed in the form of a cognitive matrix, as shown in Table 5.2.

In Table 5.2, DM represents declarative memory, that is, what students have learned; P1, P2 and P3 indicate the name of the production rule. In addition to the five modules (retrieval corresponding declarative memory module), a production module was also added in ACT-R. The content of the module production rules is activated during problem solving. Each column represents the content of a module in the problem-solving process. The leftmost column of numbers in Table 5.2 shows the line numbers, with each line representing the cognitive logic step, which is not the same as the actual solution step. The last line indicates that the cognitive process is complete; that is, the problem-solving process is complete.

(三) Cognitive simulation

ACT-R provides an abstract cognitive structure that is performed cognitively from a functional point of view. Based on the above analysis of the cognitive process of solving the "mode" problem, the Lisp program (in Appendix 2) was compiled. It was simulated in ACT-R with a minimum time interval of 0.05 s (default). The simulation of the "mode" cognitive problem-solving process is shown in Fig. 5.8.

The simulation in Fig. 5.8 shows that setting goals in the problem-solving process is a key step. It begins with a definite goal. The continuous conversion of the problem state is carried out in the middle phase, and finally, the process ends with the goal. Procedural represents procedural knowledge, that is, production rules. The extraction of procedural knowledge aims to activate production, and only one production rule can be executed at a time. Declarative represents declarative knowledge and is expressed as a chunk. The extraction of declarative knowledge occurs in the knowledge block.

The problem-solving cognitive trace can be visualized in ACT-R, and the cognitive trace of "mode" problem solving is shown in Fig. 5.9. In Fig. 5.9, the leftmost column is a buffer in ACT-R, such as a retrieval buffer, an imaginal buffer, a visual buffer, a production buffer, and a goal buffer. The right column is based on the time series (the basic event unit is the default value of 0.05 s) and offers a visual display of the contents of the buffer. The red area shows the contents of the extraction buffer, that is, the problem-solving process used to extract declarative knowledge. The yellow area shows the contents of the production buffer, which corresponds to the production of activation during problem resolution. These findings are consistent with the analysis in Table 5.2 on the cognitive process of solving the "mode" problem.

(四) Activated brain areas

The components of ACT-R map to brain regions, and this mapping can be used to record blood oxygen-dependent (BOLD) data for the "mode" problem using functional magnetic resonance imaging (fMRI) level-dependent responses.

Table 5.2 "Mode" cognitive problem-solving process analysis

	Visual	Production	Retrieval	Goal	Imaginal	Manual
1	Text encoding					
2			Relevant semantic knowledge in DM			
3				Choose which month should be chosen		
4		Choose a month when students can celebrate their birthdays, class teacher role Select ➡the birthday to celebrate that month (in line with common sense, P1)				
5				Select the month for which to celebrate birthdays (target conversion)		
6					The month with the highest number of students with birthdays	
7						The month with the highest number of students with birthdays, and then choose that month

(continued)

Table 5.2 (continued)

	Visual	Production	Retrieval	Goal	Imaginal	Manual
8				Statistics on the number of students with a birthday in each month		
9			Statistics, Count (P2)			
10					Compare the number of birthdays each month	
11			The comparison of the sizes of the numbers (P3)			
12				Choose the month in which with the largest number of students have birthdays		
13						The month in which the largest number of students have birthdays
14						End

Figure 5.10 shows the "extract buffer" after the "mode" model in ACT-R is run with BOLD data changes. The left column shows all the buffers in the "mode" problem model. If one of the buffers is selected, the blood in that buffer is displayed in the right area, and the oxygen levels depend on the data. The horizontal axis represents the time, and the default interval is 1.5 s; the vertical axis represents the range of variation, with a minimum of "0. 0" and a maximum of "1.0." Different buffers clearly correspond to different buffer zone activations.

Figure 5.11 shows the activation areas of the brain in the "mode" problem-solving process. The left column does not use the same color marked with a different buffer;

Fig. 5.8 “Mode” cognitive problem-solving process simulation

Fig. 5.9 “Mode” problem-solving cognitive traces

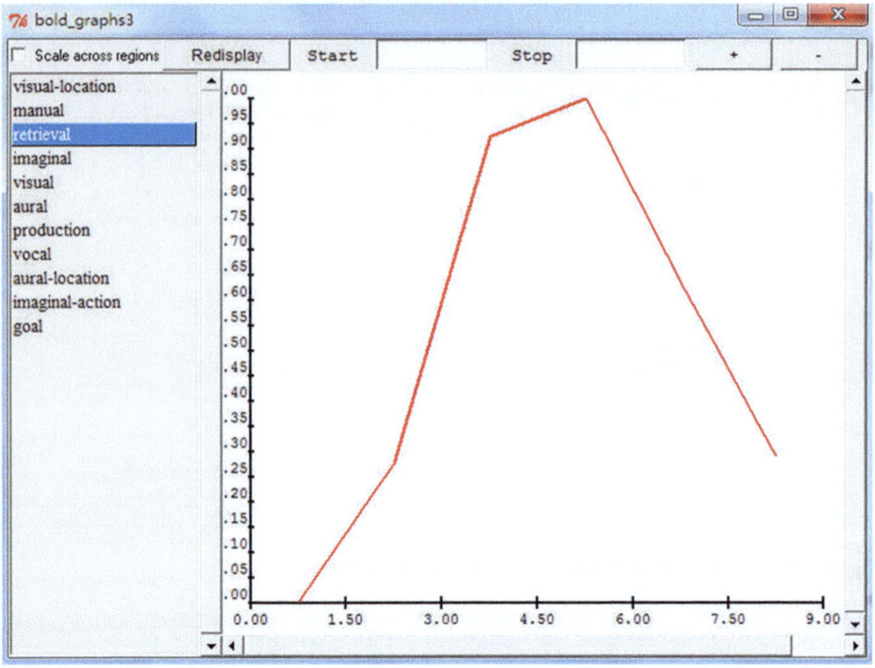

Fig. 5.10 BOLD "BOLD BUFFER" variations in the "mode" problem-solving process

the right column represents the brain activation area. The color of the area box is the same as that of the corresponding buffer on the left. The brightness of the area indicates the degree of activation. If the brightness of the area increases, the activation of the brain also increases.

Figure 5.11 clearly shows that the extraction of knowledge (mainly numbers) from the imaginal cortex is closely related to the activation of the parietal cortex. This finding is consistent with the findings of the studies by Pinel et al. (2001), Eger et al. (2003), and Zhang et al. (2004). The parietal cortex of the subjects in those studies were significantly activated when the subjects saw the numbers or performed digital processing. The default time for an image in ACT-R is 200 ms.

The retrieval buffer is responsible for extracting declarative memory and is associated with the activation of the prefrontal cortex. This finding is consistent with the findings of Qin et al. (2003), Anderson et al. (2004), and Sohn et al. (2003, 2005). The prefrontal, rather than the parietal, cortex correlates with the extraction of personal knowledge.

The procedural module is responsible for the extraction of procedural knowledge and is closely linked to the activation of the basal ganglia. This conclusion is consistent with that of Hikosaka et al. (1999).

Figure 5.12 shows the brain activation zones during the "mode" problem-solving process as a three-dimensional map. The figure "0.0–1.0" indicates the brightness

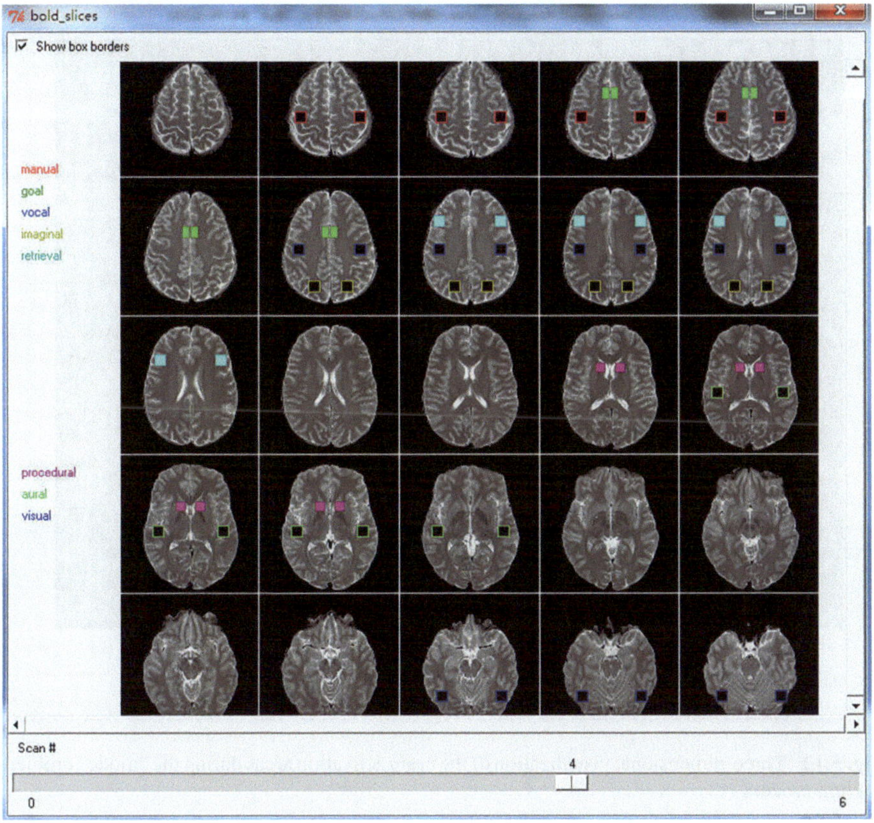

Fig. 5.11 Activation of brain areas in the "mode of" problem-solving process

value. A value of "0" means that an area has not been activated, and that area is black; the more active an area is, the closer the value is to "1," and the brighter the area is. The left side of the graph shows the buffer in a different color, and the number to the right of the buffer is the activation level. The right side of the graph is the brain activation area, shown in a color that is consistent with the color of the left module.

As shown in Fig. 5.12, in the "mode" problem-solving process, the contents of the target, extraction, and production buffers are activated. The three-dimensional visualization shows the brain activation regions corresponding to different buffers in this process. The results are the same as those in Fig. 5.11 and are supported by the existing research, which is not described here.

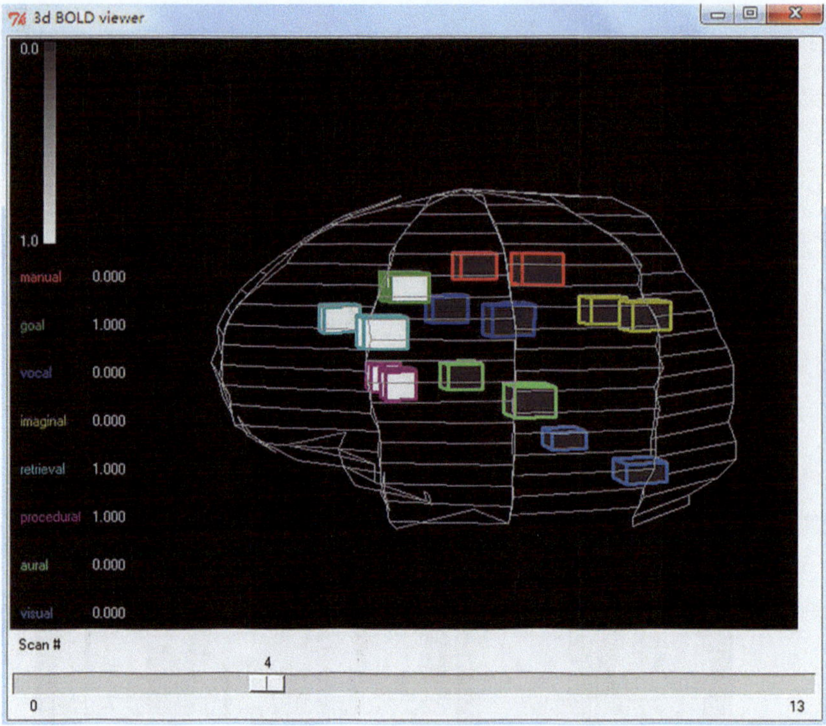

Fig. 5.12 Three-dimensional visualization of the brain activation areas during the "mode" problem-solving process

5.4 Empirical Study on the Cognitive Simulation of Procedural Knowledge Problem Solving

5.4.1 Purpose

The purpose of this experiment is to compare the consistency among procedural knowledge problem solving, cognitive process simulation and students' actual problem-solving process to verify the validity of the cognitive simulation.

5.4.2 Method

(一) The participants

A total of six students from class five, grade five at a primary school in Shijingshan District, Beijing, were selected as participants, with three boys and three girls. Two

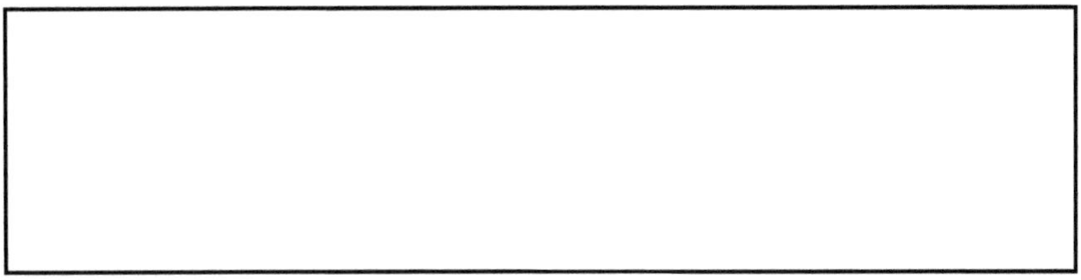

Fig. 5.13 Rectangular paper with squares

students each had excellent, middle and poor comprehensive mathematics results, with an average age of 133 months and an age range of 128–138 months.

(二) Material

The experimental materials were a few of the questions that were designed according to the purpose of this book.

(1) In grade five, class two, the skipping test was carried out. The 1-min rope skipping results for the first group of seven students are as follows:

172 145 135 142 139 140 138

What number do you think would indicate the average level of jumping rope for this group of students?

(2) Please color the rectangular paper (as shown in Fig. 5.13). Color 1/3 of the paper yellow and 2/5 of the paper black, and do not use the same color for another region (the yellow area cannot be colored black, and the black area cannot be colored yellow). How much of the whole piece of paper is yellow and black?

The first question aimed to train students to think aloud when solving a problem, and the second question tested students' knowledge of "addition with different denominators."

(三) Program

(1) Design the experimental scheme

According to the task and purpose of the research, the subjects, materials and instructions were determined, and the oral report records were analyzed. Then, the records were compared with the cognitive simulation results.

(2) Experimental equipment

A Sony's recording pen, Sony's digital video camera and tripod were used to record the oral reports during the experiment.

(3) Oral reporting and recording

The oral reporting method was applied to collect information. In accordance with the think aloud research program developed by Erickson and Simon, the subjects were trained to think aloud in the process of solving the problem. The instructions were as follows: "Please read the questions aloud, think about the process of solving the problem, and speak what you think. In other words, speak while completing the questions; speak your thinking process aloud, so you know how you perform it." Before students began answering the question, the main tester (the researcher himself) briefly explained the instructions. Then, taking question (1) as an example, the main tester demonstrated and explained how to think aloud in the process of solving the questions. After the subjects understood the instruction to think aloud, they began to answer the question. A camera was used to record the students' problem-solving process.

(4) Data translation and coding

The data collected included the two parts of the oral report and problem-solving operations. The oral report was first translated into text by the experts, and then the students' problem-solving assignments were encoded and analyzed to diagnose their answers. Two experts were responsible for coding, and agreement was reached after a discussion of a small number of coding inconsistencies.

Simon et al. note that the intuitive information provided in oral reports concerns the knowledge and information needed to solve a problem, not the actual process used (Simon and Newell 1989). Therefore, it is necessary to deduce the process from the oral reports rather than trying to encode the process directly.

5.4.3 Results Analysis

Newell and Simon implemented computer simulations of human thinking and inferred the validity of the simulations by comparing them with spoken language reports (Newell & Simon, 1961). Based on the research foundation, this experiment compares the simulation process with the students' oral report records to determine the validity of the simulation.

(一) Oral report analysis

Table 5.3 provides a detailed description of the process of solving the problem of addition with different denominators.

The content of (·) is omitted in the oral reports of students. To indicate completion, it is added, along with a (·) mark.

An analysis of the "addition with different denominators" oral reports reveals that WangZY, ChenHY and XingYR and other students solved the problem of addition with different denominators, including the reduction of fractions to a common denominator, the least common multiple, and other steps. However, in the least common multiple step, WangZY mentioned that "3 and 5 are coprime, and the least common multiple is $3 \times 5 = 15$," and ChenHY and XingYR directly stated that "the

Table 5.3 Oral report and cognitive process analysis of "addition with different denominators"

The subjects	Oral report	Cognitive process analysis
	[Read the question] Please color the rectangular paper. Color 1/3 of the paper yellow and 2/5 of the paper black. The same color cannot be used for another area (the yellow area cannot be colored black, and the black area cannot be colored yellow). How much of the whole piece of paper is yellow and black?	Input text information by reading the title, and form the propositional text frame and the problem pattern after visual coding
(Student WangZY)	[Analysis] The denominators of the fractions (two fractions 1/3, 2/5) are 3 and 5, both of which are coprime, with a least common multiple of $3 \times 5 = 15$. Then, $1 \times 5 = 5$, $2 \times 3 = 6$, that is, 5/15, 6/15, $5/15 + 6/15 = 11/15$	Activation of long-term declarative memory related to the concept of "mutual prime"; activation of production rule P1: prime multiples of the least common multiple → Multiply two numbers together, $3 \times 5 = 15$; Activate production rule P2: Determine the LCM → the numerator and denominator of 1/3 are multiplied by the same number 5, and the numerator and denominator of 2/5 are multiplied by 3 Activation of production rule P3: Add with the common denominator → The denominator does not change, and the numerators are summed
(Student ChenHY)	[Read the question] Please color the rectangular paper. Color 1/3 of the paper yellow and 2/5 of the paper black. The same color cannot be used for another area (the yellow area cannot be colored black, the black area cannot be colored yellow). How much of the whole piece of paper is yellow and black?	Input text information by reading the title, and form the propositional text frame and the problem pattern after visual coding

(continued)

Table 5.3 (continued)

The subjects	Oral report	Cognitive process analysis
	[Analysis] Reduction of fractions to a common denominator. The least common multiple of 3 and 5 is 15, 1/3 × 5 = 5 / 15, 2/5 × 3 = 6/15 The numerators add up to 11	Activation of production rule P1: Addition of denominator fractions → reduction of fractions to a common denominator; activate "reduction of fractions to a common denominator." Production rule P2: reduction of fractions to a common denominator → calculate the least common multiple of the denominator; Production rule P3: 3 and 5 are prime numbers, find the least common multiple → the least common multiple is 3 × 5 = 15; Production rule P4: the least common multiple of 1/3 and 2/5 is 15, translate into the same denominator → the denominator is converted to the least common multiple of 15. The denominator and the numerator are multiplied by the same number; the numerator and denominator of 1/3 are multiplied by the same number, 5, resulting in 5/15; the numerator and denominator of 2/5 are multiplied by 3, resulting in 6/15 Activation of production rule P4: Add fractions with the common denominator → The denominator does not change, and the numerators are summed
	[Read the question] Please color the rectangular paper. Color 1/3 of the paper yellow and 2/5 of the paper black. The same color cannot be used for another area (the yellow area cannot be colored black, the black area cannot be colored yellow). How much of the whole piece of paper is yellow and black?	Input text information by reading the title, and form the propositional text frame and the problem pattern after visual coding

(continued)

Table 5.3 (continued)

The subjects	Oral report	Cognitive process analysis
(Student XingYR)	[Analysis] Determination of the percentage of yellow and black to the whole piece of paper, reduction of fractions to a common denominator, conversion of fractions to the same denominator of 15 Convert 1/3 into 5/15, and turn 2/5 into 6/15; 6/15 + 5/15 = 11/15. This method feels a little messy. This paper can be divided into 15 parts [Ask: why is it divided into 15 parts?] Because the denominator is 15, and the yellow and black together accounted for the whole piece of paper. The denominator is 15 after the reduction of fractions to a common denominator. The whole paper is divided into 15 parts, 11 parts are selected from it, and yellow and black together accounted for 11/15 of the sheet of paper	(1)Determine the target Activation of production rule P1: Addition of denominator fractions → reduction of fractions to a common denominator; activate "reduction of fractions to a common denominator." Production rule P2: reduction of fractions to a common denominator → calculate the least common multiple of the denominators; Production rule P3: 3 and 5 are prime numbers, find the least common multiple → the least common multiple is $3 \times 5 = 15$; Production rule P4: the least common multiple of 1/3 and 2/5 is 15, translate the denominators into the same denominator → the denominator is converted to the least common multiple of 15. The denominators and the numerators are multiplied by the same number; the numerator and denominator of 1/3 multiplied by the same number, 5, resulting in 5/15; the numerator and denominator of 2/5 are multiplied by 3, resulting in 6/15. Activation of production rule P5: Add fractions with the common denominator → The denominator does not change, and the numerators are summed (2) Reflect the problem-solving process

(continued)

Table 5.3 (continued)

The subjects	Oral report	Cognitive process analysis
	[Read the question] Please color the rectangular paper. Color 1/3 of the paper yellow, 2/5 of the paper black, and the same color cannot be used for another area (the yellow area cannot be colored black, the black area cannot be colored yellow). How much of the whole piece of paper is yellow and black?	Input text information by reading the title, and form the propositional text frame and the problem pattern after visual coding
(Student LiL)	[Analysis] The denominator should be 15, (divide the rectangle into three in the figure, and paint one of them). The approach is to divide the remaining paper into 5 parts and take 2 of them (2/5 of the whole paper). Yellow and black together accounted for a fraction of the entire piece of paper, that is, 1/3 + 2/5; the denominator is 15, and the numerator is 2, resulting in 2/15 [Ask: 2 how did you figure it out?] 2 is the numerator: 1 × 2, following the least common multiple (referring to 15), the operation is 3 × 5, and the denominator is calculated accordingly	The student understands that 2/5 of the whole piece of paper is wrong, and he understands that 2/5 represents the remainder Activation of the wrong production rule P1: Addition with different denominators → Denominator and numerator multiplied by respective numbers Further confirmed the activation of production rule P1: addition with different denominators → denominator and numerator are multiplied

least common multiple is $3 \times 5 = 15$." LiL incorrectly solved this problem because of the use of the incorrect production rule.

(二) Cognitive simulation and oral report comparison.

Figure 5.14 shows a comparison of "addition with different denominators" problem-solving cognitive simulation and oral reports. The left panel is the result of the simulation, and the right panel is the content of the oral report. The comparison shows agreement between the two.

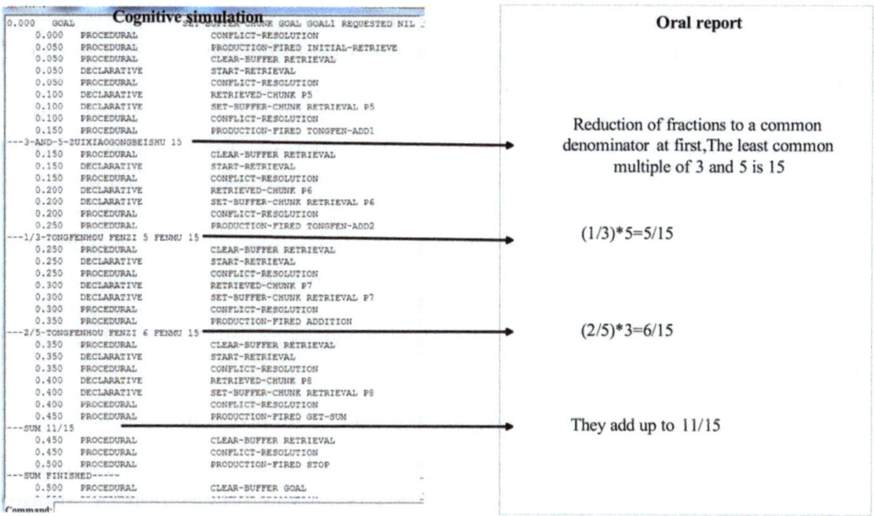

Fig. 5.14 Cognitive simulation and oral report comparison for "Addition with different denominators"

5.4.4 Discussion

(一) On the same question, different students used different methods for solving the same problem.

With respect to the problem of "addition with different denominators," WangZY, ChenHY and XingYR all solved the problem correctly, but the details of their approaches were different. In seeking the least common multiple, WangZY mentioned the concept that "3 and 5 were coprime, and the least common multiple was $3 \times 5 = 15$," which activated the concept of "coprime" in long-term declarative memory. Specifically, to find the least common multiple, according to the qualities of the mutual prime, the least common multiple is multiplied by two; this reasoning was activated in long-term program memory. In contrast, ChenHY and XingYR directly stated that "the least common multiple is $3 \times 5 = 15$," which activated long-term programmatic memory.

(二) In the process of solving the problem, students showed different degrees of "automation."

In the "addition with different denominators" problem, to solve "the least common multiple of 3 and 5," WangZY's approach was "3 and 5 are coprime, and the LCM is $3 \times 5 = 15$." ChenHY's step was the direct determination of the "least common multiple of 3 and 5 is 15," which provided the result. This example shows that in the process of solving problems, the internal operation can be compressed, and after a long period of training, a few simple internal operations may be compressed into one "chunk." For example, the two production rules P1: $A \rightarrow B$ and P2:$B \rightarrow C$ are often activated at the same time, which generates a new production rule P3:A → C. Additionally, the phenomenon of "speeding up" occurred when this student studied algebraic equations. He believed that after training, the solution of the equation could be simplified into a series of visual coding and output operations (Anderson, 2005). The study by Schoenfeld shows that becoming an expert in a field generally requires approximately 50,000 knowledge blocks in long-term memory, which are concrete objects of thinking in the field; moreover, in many cases, the use of strategy is, in fact, the use of such a well-established knowledge block (Schoenfeld, 1985). The above conclusions are in line with the analysis of this study, which also explains, to some extent, the difference between experts and novices in solving complex problems. Experts have more knowledge of "automation," whereas few beginners have this knowledge.

(三) Incorrect production is an important reason for the problem-solving error

In the problem of "addition with different denominators," when LiL solves "1/3 + 2/5," the incorrect production Equation P1 is activated: denominator addition, denominator multiplication and numerator multiplication, resulting in a problem-solving error. There are two reasons for this error. First, LiL does not understand the meaning of the fraction. There is a problem with semantic models of fractions in long-term declarative memory. Second, he does not understand the tactics of reducing fractions to a common denominator through the divisions mentioned earlier. In addition, the reason for the reduction of fractions to a common denominator and how to do so are not known. Anderson studied the cognitive process of students learning solutions to algebraic equations and suggested that learning occurs at the symbolic level and creates (or generates) new production rules (Anderson, 2005). Therefore, helping students form the correct production rules is an important part of procedural knowledge learning.

(四) Cognitive analysis of problem-solving process helps diagnose and intervene in problems

LiL makes a typical error in calculating the "addition of denominators," as indicated by the oral report: (1) LiL successfully extracted the declarative knowledge $3 \times 5 = 15$ and $1 \times 2 = 2$, which indicated that there was no problem in multiplying two numbers. (2) Although the numerator and denominator were multiplied, the report showed that he could correctly identify the numerator and the denominator of the fraction. (3) The error in solving the problem stems from the use of incorrect production rules: "the addition of denominators → the numerator and the denominator are multiplied separately, and the numerators are summed. To help LiL correct his mistakes, we must consider how to help him form the correct production rule of "addition with different denominators → find the least common multiple" and the basic operation needed to achieve this production rule.

(五) Determine whether cognitive simulation is consistent with the student's problem-solving process

Whether a computer can completely simulate the human problem-solving process has always been disputed. The task of thinking aloud, which was proposed by Newell and Simon (1961), effectively answered this question and promoted the development of cognitive psychology. It provided a new perspective on the study of human thinking and later developed into an important method in psychology research—the oral report method.

Because of existing knowledge, learning styles, cognitive characteristics, the family environment, and other factors, students will not be consistent in their answers to the same questions, but there will always be similarities. As Newell and Simon demonstrated in computer simulations through a verbal reporting approach, not all individuals adopt the same problem-solving process, but many similarities and commonalities exist. In this book, we consider mainly the commonalities.

5.5 Declarative Knowledge Problem-Solving Cognitive Simulation Empirical Research

5.5.1 Purpose

The purpose of the experiment is to verify the validity of the cognitive simulations by comparing the cognitive simulation of the declarative knowledge problem-solving process with the students' practical problem-solving process.

5.5.2 Method

(一) The subjects

A total of six students from grade five, class three at a primary school in Shijingshan District, Beijing, were selected as subjects, comprising three males and three females. Two students each had excellent, middle and poor students' comprehensive mathematics results, with an average age of 133 months and an age range of 131–135 months.

(二) Material

The experimental materials are a few of the questions that were designed according to the purpose of this book.

(1) In grade 5, class two, the rope skipping test was carried out. The results for the first group of seven students for 1 min of rope skipping are as follows:

172 145 135 142 139 140 138

What number do you think is suitable for indicating the average level of rope skipping for this group of students?

(2) The school has agreed to hold a birthday celebration next year for grade five, class three. However, only the birthdays of students born in a certain month can be celebrated. If you are the class teacher:

① How would you choose the month?
② Which month do you think should be chosen?

The first question was used to train students in think aloud exercises, and the second question was designed to test students' knowledge of "the mode" topic.

(三) Program

(1) Design experiment scheme

According to the task and purpose of the research, the subjects, materials, and instructions were determined, the oral report records were analyzed, and the results were compared with the cognitive simulation results.

(2) Experimental equipment

A Sony's recording pen, Sony's digital video camera and tripod were used to record the oral reports during the experiment.

(3) Oral reporting and recording

The oral reporting method was used to collect information. In accordance with the think aloud research program developed by Erickson and Simon (1981), the subjects were trained to think aloud while solving a problem. The instructions were as follows: "Please read the questions aloud, think about the process of solving the problem, and say what you think. In other words, in the process of problem-solving, speak your thinking process aloud so that you know how you perform it." Before the students began answering the question, the main tester (the researcher himself) first briefly explained the instructions. Then, taking question (1) as an example, the main tester demonstrated and explained how to think aloud in the process of completing the questions. After the subjects understood how to think aloud, they began to answer the question, and the students' problem-solving process was recorded.

(4) Data translation and coding

The data collected included two parts: oral reports and problem-solving operations. The oral report was first translated by professionals into text, and then the students' problem-solving assignments were encoded and analyzed to diagnose issues in the problem-solving process. Two experts were responsible for coding work, and agreement was reached after a discussion of a small number of coding inconsistencies.

Simon et al. note that the intuitive information usually provided in oral reports is about the knowledge and information needed to solve a problem, not the actual process used (Simon & Kaplan, 1989). Therefore, it is necessary to deduce the process from the oral reports rather than trying to encode the process directly.

5.5.3 Results Analysis

Newell and Simon (1961) implemented computer simulations of human thinking and inferred the validity of the simulations by comparing them with oral reports. Based on this research foundation, this experiment compared the simulation process with students' oral report records to verify the validity of the simulation.

(一) Oral report analysis

Table 5.4 provides a detailed description of the process of solving the "mode" problem.

The content of (·) is omitted in the oral reports of students. To indicate completion, it is added, along with a (·) mark.

(二) Comparison of cognitive simulation and oral report

Figure 5.15 shows a comparison of the cognitive simulation for the "mode" problem-solving process and the oral reports. The left panel shows the results of the simulation, and the right panel shows the content of the oral report. The comparison shows that the two are the same.

5.5.4 Discussion

(一) For the same question, different students provide different details concerning the problem, but they all have commonalities.

With respect to the "mode" problem, students QiuDL and LiC solved the problem correctly. By analyzing the oral reports of the two students, we found that in counting how many students had a birthday each month, the two students adopted different methods. QiuDL first listed the month of each person's birthday and then determined which month had the most birthdays. LiC drew a table from January to December and then counted the number of January classmates who stood up; then, the process was repeated for students who were born in the remaining months. Finally, the correct month was selected. QiuDL and LiC went through the process of identifying the problem (finding the month with the highest number of birthdays), counting the number of birthdays in each month, comparing the number of birthdays each month, and determining the number of months with the highest number of birthdays, all of which were common to both students' processes. As Newell and Simon showed through colloquial reporting to validate computer simulations, not everyone's problem-solving process is the same, but there are many similarities and commonalities (Newell & Simon, 1961). In this book, we consider mainly the commonalities.

(二) Determining a goal is a key aspect of successful problem solving

Table 5.4 Oral report and cognitive process analysis of the "mode"

The subjects	Oral report	Cognitive process analysis
	[Read the question] The school agrees to hold a birthday celebration next year for class three, grade five. However, it can be celebrated only for students born in a certain month. If you are the class teacher: ① How would you choose the month? ② Which month do you think should be chosen?	Input text information by reading the title, and form the propositional text frame and the problem pattern after visual coding
(Student QiuDL)	[Analysis] (Thinking in 49 s) I do not know the birthday of everyone in our class. We choose the month in which most students have birthdays. This is the question, specifically, which month? First, I think the month of every student should be listed, and then which month has the most birthdays should be found and then chosen. Most students' birthdays can be taken care of	Identify the problem: the goal is to find the month in which the most students have a birthday Statistics on the dates of students' birthdays Choose the month with the largest number of birthdays The month in which the most students have birthdays should be selected Understand the problem situation, taking into account the role of the head teacher
	[Read the question] The school agrees to hold a birthday celebration next year for class three, grade five. However, it can be celebrated only for students born in a certain month. If you are the class teacher: ① How would you choose the month? ② Which month do you think should be chosen?	Inputting text information by reading the title, and form the propositional text frame and the problem pattern after visual coding
(Student LiC)	[Analysis] The month in which most students have a birthday should be selected. There should be statistics on the month in which students have birthdays Choose the month with the largest number of birthdays [Asked: How can the statistics be generated?] Draw a table of January to December, (then) count the number of January classmates who stand up, then repeat the process for students who are born in February. Finally, make sure the correct month is selected	Identify the problem. The goal is the month in which students have birthdays Choose a solution strategy Activate procedural memory "statistics" production Activate "comparative" production in procedural memory

(continued)

Table 5.4 (continued)

The subjects	Oral report	Cognitive process analysis
	[Read the question] The school agrees to hold a birthday celebration next year for class three, grade five. However, it can be celebrated only for students born in a certain month. If you are the class teacher: ① How would you choose the month? ② Which month do you think should be chosen?	Input text information by reading the title, and form the propositional text frame and the problem pattern after the visual coding
(Student ChenYL)	[Analysis] I choose May [Asked: How did you choose May?] There were more May birthdays [Asked: How do you know that the most students celebrate their birthday in May?] Ask classmates. Most students have their birthday in May. Each year the school restaurant gave birth birthday cake hair, and then you can know the students in which month birthday	Extract long-term declarative memories for "the largest number of birthdays is May." (The implicit assumption is that the goal is to select the month with the largest number of birthdates.) Choose a problem-solving strategy Activate the long-term declarative memory of "school birthday" scene The birthday cake will be sent to the student → The month of sending a cake is the month with the largest number of birthdays
	[Read the question] The school agrees to hold a birthday celebration next year for class three, grade five. However, it can be celebrated only for students born in a certain month. If you are the class teacher: ① How would you choose the month? ② Which month do you think should be chosen?	Input text information by reading the title, and form the propositional text frame and the problem pattern after visual coding
(Student PangB)	[Analysis] I think the months of January, March, May, July, August, October and December are most appropriate because they all have 31 days in a month. More classmates will have birthdays in these months January is the winter vacation, and July and August are the summer vacation, so choose between March and May. Choose May There are more activities in October because there are 7 national days, and the time of one's birthday is particularly tense Choose between March and May. I feel that March is the best because in the middle of the semester, there is not too much pressure, so the celebration will not waste time for learning. I feel the March is best and most suitable month	Activate production rule P1: Which month has more days → There are more people who have a birthday, and the assumptions are: Everyone's birthday is evenly distributed on a daily basis, so the more days in a month, the more people that have birthdays Errors are produced, which lead to solving the incorrect problem

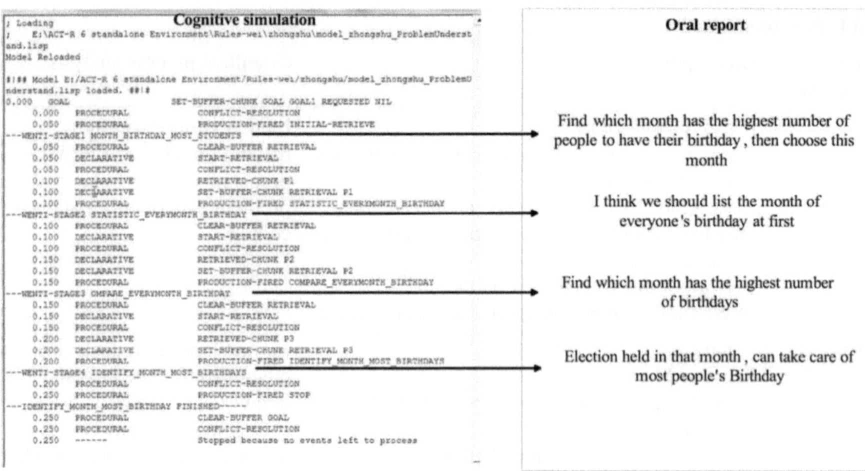

Fig. 5.15 Comparison of the "mode" cognitive simulation and oral report

An analysis of the "mode" oral report revealed that in the problem-solving processes adopted by QiuDL and LiC, the first step was correctly determining the target and then adopting a strategy to solve the problem successfully. Although ChenYL decided that the goal was to select the months in which the most students who had birthdays, he did not consider the whole class, which led to a problem-solving error.

(三) Understanding the problem situation an important step in successfully solving the problem

"Mode" is a typical type of declarative knowledge, and the process of obtaining the concept of "mode" depends largely on students' experiences. The phrase "if you are the class teacher" gives students the "class teacher" role; students moreover must consider a "birthday" situation and, to gain a deeper understanding of the "class teacher" role, will take into account the fairness of accounting for the most students and then select the month with the most birthdays. Students QiuDL and LiC have a good understanding of this situation and set the problem goal accordingly. When ChenYL answered the question, he activated the situation of "birthday cake for the student at school" and quickly determined the month when there were the most cakes, that is, the month with the most birthdays. The teacher confirmed that ChenYL is a student at school and personally experienced a cake scene. However, ChenYL only considered the birthday of a student at school and did not consider the birthday of the whole class. That is, he did not understand the role of the "teacher in class" thoroughly, resulting in problem-solving errors.

(四) In the process of solving the problem, there are different degrees of "automation."

ChenYL directly answers the question when solving the "mode" problem. An analysis of the oral report revealed that although there are some errors in the process of solving problems and the determination of goals and other stages, the activation of long-term declarative memory of students regarding the months with the most birthdays is May. By directly stating the answer is May, a few simple operations are combined into a "chunk," resulting in "automation." Anderson studied the phenomenon of "speeding up" in his study of algebraic equations. He believed that after enough training, the solution equation could be simplified into a series of visual coding and output operations (Anderson, 2005). This view is consistent with the analysis of this book.

(五) Incorrect production is an important reason for the problem-solving error

In the process of solving the "mode" problem, Pang B activated the wrong production "which month has more days → there are more people who have a birthday" when he selected the month with the most birthdays, leading to problem-solving errors. With respect to the role of production in learning, Anderson studied the cognitive process of students learning solutions to algebraic equations and suggested that learning occurs at the symbolic level, creating (or generating) new production rules (Anderson, 2005). Therefore, helping students form the correct production rules is an important part of procedural knowledge learning.

5.6 Cognitive Simulation Contributions and Limitations

The computer simulation of problem solving has an important influence on the development of artificial intelligence. It has promoted the in-depth study of the psychology of problem solving and increased people's understanding of certain aspects of problem solving.

(一) Computer simulation helps solve the problem of visualizing internal cognitive processes. Psychological studies of problem solving give more attention to one of the links, such as problem characterization, strategy selection, etc. Thus, the entire process cannot be visualized. In recent years, cognitive neuroscience research on problem solving has made many achievements and provided some evidence. However, these results are more concentrated at the nervous system level but cannot be used to characterize the internal process explicitly. Computer simulation involves the entire process of problem solving and visually displays this internal cognitive process, clearly showing the procedural and declarative knowledge required for problem solving.

(二) Computer simulations have promoted the organization of research into knowledge bases. Computer simulation of problem solving requires a knowledge base as a support. The more knowledge in a knowledge base, the easier the information processing is and the more easily a problem can be solved. Common knowledge in the knowledge base will form a large chunk in long-term memory, which provides ideas for the decomposition and combination of knowledge in mathematics teaching.

(三) Computer simulation presents the concept of a production system. Production systems formalize the cognitive activities of problem solving. The rules of "if (conditions) then (action)" apply to different content, and different kinds of problems become the general mechanism of problem solving. The productive formula emphasizes the importance of correct identification in problem as well as the correct application of the premise. Therefore, the problem-solving process becomes the process of obtaining and applying the production system correctly. The production system supplies new ideas for solving mathematical problems.

Although some achievements have been made in computer simulations of problem solving, some problems still need further study.

(一) Problem-solving computer simulation programs are performed in a serial fashion, yet individuals' thinking processes when solving problems may not be serialized. This issue is also very controversial in psychology.

(二) When a problem is solved, individuals provide a quick response according to the situation at that time, which has a certain degree of randomness. However, computer simulations cannot take into account the situation.

(三) Computer simulations of problem solving do not consider motivations, emotions, attitudes or other factors in the problem-solving process. These factors have a significant impact on problem solving and can play a role in selecting, guiding and controlling cognitive processes.

Although many problems remain in the problem-solving computer simulation, computer programs operate according to strict logic and certainty. The cognitive process of problem solving cannot be accomplished by other means. Computer simulation combines some of the factors in the process of problem solving to reconstruct this process and overcomes the formerly analytical approach of experimental psychology. This advancement opens a path for understanding the cognitive process of problem solving as a whole (Wang & Wang, 1992). Therefore, it is a special research method of cognitive psychology and is highly important for computer simulation.

5.7 Summary

This chapter first describes the basis of problem-solving cognitive simulation and then introduces the cognitive model used as a tool, namely, adaptive control of thought-rational (ACT-R), along with the internal structure, application areas and cognitive neurology. We subsequently selected procedural knowledge problems ("addition with different denominators") and declarative knowledge problems ("mode"), analyzed the cognitive process of problem solving, and used Lisp to write cognitive programs. In ACT-R, cognitive simulation was performed, the results were visualized, and the brain regions activated during problem solving were analyzed. To verify the validity of the cognitive simulation, a group of students from two classes in a primary school were tested via oral English reports. The results showed that the cognitive simulations were consistent with the oral reports.

The computer simulation of problem solving has an important influence on the development of artificial intelligence. It promotes the in-depth study of psychology in problem solving and improves individuals' understanding of certain aspects of problem solving. However, many problems remain in the current computer simulation problem-solving process. Cognitive simulation reveals that the cognitive process of problem solving is irreplaceable by other means. This insight opens the way for understanding the cognitive process of problem solving as a whole.

The problem-solving cognitive process is analyzed and simulated to visualize the implicit process of problem solving. This approach not only helps to deepen the understanding of cognitive processes but also helps diagnose students with learning disabilities. It provides targeted counseling to help improve academic performance.

Appendix 1

file name: "model_yifenmuixangjia.lisp"

```
(clear-all)
(define-model yifenmuxiangjia
(sgp :esc t :lf .05)
(chunk-type known-condition object value)
(chunk-type fenshu object fenzi fenmu)
(chunk-type zuixiaogongbeishu fenmu1 fenmu2 value)
(chunk-type tongfen zuixiaogongbeishu fenshu   value)
(chunk-type addition addend1   addend2   sum stage)
(chunk-type count-tongfenmu tongfenmu1 tongfenmu2   sum)
(add-dm
    (yellow-grid isa chunk)
    (black-grid isa chunk)
    (qiuzuixiaogongbeishu isa chunk)
    (tongfen-add1 isa chunk)
    (tongfen-add2 isa chunk)
    (addition isa chunk)
    (p1 ISA known-condition object yellow-grid   value 1/3)
    (p2 ISA known-condition object black-grid   value 2/5)
    (p3 ISA fenshu object 1/3 fenzi 1 fenmu 3)
    (p4 ISA fenshu object 2/5 fenzi 2 fenmu 5)
    (p5 ISA zuixiaogongbeishu fenmu1 3 fenmu2 5 value 15)
    (p6 ISA tongfen zuixiaogongbeishu 15 fenshu 1/3 value 5/15)
    (p7 ISA tongfen zuixiaogongbeishu 15 fenshu 2/5 value 6/15)
    (p8 ISA count-tongfenmu tongfenmu1 5/15 tongfenmu2 6/15 sum 11/15)
    (goal1 ISA addition   addend1 1/3 addend2 2/5 sum nil stage nil)
)
(p initial-retrieve
    =goal>
        ISA         addition
        addend1     =add1
        addend2     =add2
sum             nil
stage           nil
  ==>
    =goal>
stage       qiuzuixiaogongbeishu
    +retrieval>
```

```
              ISA           zuixiaogongbeishu
    ;;!output! (---zuixiaogongbeishu =num1)
    )
    (p tongfen-add1
         =goal>
              ISA           addition
              addend1       =add1
              addend2       =add2
      stage        qiuzuixiaogongbeishu
           =retrieval>
               ISA          zuixiaogongbeishu
      value        =val
        ==>
            =goal>
      stage     tongfen-add1
           +retrieval>
               ISA                         tongfen
      fenshu                         =add1

   !output! (---3-and-5-zuixiaogongbeishu =val)     ;;输出 3 和 5 的最小公倍数为 15
    )
    (p tongfen-add2
         =goal>
              ISA           addition
              addend1       =add1
              addend2       =add2
      stage        tongfen-add1
           =retrieval>
               ISA                         tongfen
      zuixiaogongbeishu        =num
      fenshu                         =add1
      value                          =tongfenhou-add1
       ==>
            =goal>
      stage     tongfen-add2
           +retrieval>
               ISA                         tongfen
      zuixiaogongbeishu        =num
      fenshu                         =add2
      !output! (---1/3-tongfenhou fenzi 5 fenmu 15);;!output! (---1/3-tongfenhou   =tongfenhou-
```

```
add1)        ;;输出 1/3 通分后的结果
    )
    (p addition
        =goal>
            ISA          addition
            addend1      =add1
            addend2      =add2
    stage        tongfen-add2
        =retrieval>
            ISA                    tongfen
    zuixiaogongbeishu      =num
    fenshu                 =add2
    value                  =tongfenhou-add2
    ==>
        =goal>
    stage     addition
        +retrieval>
            ISA                count-tongfenmu
    !output! (---2/5-tongfenhou fenzi 6 fenmu 15);;!output! (---2/5-tongfenhou =tongfenhou-add2)        ;;输出 2/5 通分后的结果
   )
  (p get-sum
        =goal>
            ISA          addition
            addend1      =add1
            addend2      =add2
    stage        addition
        =retrieval>
            ISA                count-tongfenmu
    sum                =count
    ==>
        =goal>
    sum        =count
     !output! (---sum =count)
    )
    (P stop
        =goal>
            ISA          addition
            addend1      =add1
            addend2      =add2
```

```
sum            =count
==>
      !output! (---sum finished-----)
    -goal>
)
(goal-focus goal1)
)
```

Appendix 2

file name: "model_zhongshu_ProblemUnderstand.lisp"

```
(clear-all)
(define-model zhongshu
(sgp :esc t :lf .05)
(chunk-type role-affair role affair value)
(chunk-type selected_month value stage)
(chunk-type statistics category amount)
(chunk-type compare category amount value)
(add-dm
    (banzhuren isa chunk)
(select-month isa chunk)
  (month_birthday_most_students isa chunk)
  (statistic_everymonth_birthday isa chunk)
  (compare_everymonth_birthday isa chunk)
    (identify_month_most_birthday isa chunk)
    (yuefen isa chunk)
    (renshu isa chunk)
    (month-has-largest-number isa chunk)
   (p1 ISA role-affair role banzhuren affair select-month value
month_birthday_most_students)
   (p2 ISA statistics category yuefen amount renshu)
   (p3 ISA compare category yuefen amount renshu value month-has-largest-number)
   (goal1 ISA selected_month value nil stage nil)
)
(p initial-retrieve
   =goal>
      ISA selected_month
value nil
  ==>
    =goal>                    ;;目标转换，过生日最多的月份
stage month_birthday_most_students
    +retrieval>
      ISA role-affair
role banzhuren
!output! (---wenti-stage1 month_birthday_most_students)
;;输出将目标转换求为过生日最多的月份
)
  ;;统计每个月过生日的人数
 (p statistic_everymonth_birthday
```

```
    =goal>
       ISA          selected_month
stage        month_birthday_most_students

  ==>
    =goal>
stage    statistic_everymonth_birthday
    +retrieval>
      ISA          statistics
category     yuefen
 !output! (---wenti-stage2 statistic_everymonth_birthday)
;;输出统计每个月过生日的人数
 )
(p compare_everymonth_birthday     ;;比较每个月过生日的人数
    =goal>
        ISA          selected_month
stage        statistic_everymonth_birthday

  ==>
    =goal>
stage        compare_everymonth_birthday
    +retrieval>
      ISA          compare
category     yuefen
 !output! (---wenti-stage3 ompare_everymonth_birthday)
;;输出比较每个月过生日的人数
 )

 (p identify_month_most_birthdays                  ;;确定人数最多的月份
    =goal>
        ISA          selected_month
stage        compare_everymonth_birthday
     ==>
    =goal>
stage        identify_month_most_birthday
!output! (---wenti-stage4 identify_month_most_birthdays)
;;输出确定人数最多的月份
 )

(P stop
```

```
        =goal>
            ISA          selected_month
     stage       identify_month_most_birthday
  ==>
        !output! (---identify_month_most_birthday finished-----)
        -goal>
     )
     (goal-focus goal1)
 )
```

References

Anderson, J. R., Qin, Y., Stenger, V. A., & Carter, C. S. (2004). The relationship of three cortical registries to an information to processing model. *Cognitive Neuroscience, 16*, 637–653.

Anderson, J. R., Fincham, J. M., Qin, Y. L., & Stocco, A. (2008). A central circuit of the mind. *Trends in Cognitive Sciences, 1*, 136–143.

Anderson, J. R. (2005). Human symbol manipulation within an integrated cognitive architecture. *Cognitive Science, 29*, 313–341.

Ashby, F. G. G., & Waldron, E. M. (2000). The neuropsychological bases of category learning. *Current Programs in Psychological Science, 9*, 10–14.

Baddeley, A. D. (1986). *Working memory* (pp. 57). Oxford University Press.

D'Esposito, M., Piazza, M., Detre, J. A., Alsop, D. C., Shin, R. K., Atlas, S., et al. (1995). The neural basis of the central executive of working memory. *Nature, 378*, 279–281.

Eger, E., Sterzer, P., Russ, M. O., Giraud, A - L., & Kleinschmidt, A. (2003). A supramodal number representation in human intraparietal cortex. *Neuron*, 37, 719–726.

Gelernter, H., Hansen, J. R., & Loveland, D. W. (1960). *Empirical explorations of the geometry. In theorem machine* (pp. 143–150). Western Joint IRE-AIEE-ACM Computer Conference.

Hikosaka, O., Nakahara, H., Rand, M. K., Sakai, K., Lu, Z., Nakamura, K., et al. (1999). Parallel neural networks for learning sequential procedures. *Trends in Neuroscience, 22*, 464–471.

Hiller, L. A., & Isaacson, L. M. (1959). *Experimental music* (pp. 17). McGraw-Hill.

Newell, A., & Simon, H. A. (1961). Computer simulation of human thinking. *Science, 134*, 2011–2017.

Newell, A., Shaw, J. C., & Simon, H. A. (1957). *Programming the logic theory machine* (pp. 218–230). ACM.

Newell, A., Shaw, J. C., & Simon, H. A. (1958). Chess-growing programs and the problem of complexity. *IBM Journal of Research and Development, 2*, 320–335.

Newell, A., & Simon, H. A. (1956). The logic theory machine-A complex information processing system. *IRE Transactions on Information Theory, 2*, 61–79.

Newell, A., Shaw, J. C., & Simon, H. A. (1959). *Report on a general problem-solving program* (pp. 256–264). IFIP Congress.

Pinel, P., Dehaene, S., Riviere, D., & Le Bihan, D. (2001). Modulation of parietal activation by semantic distance in a number comparision task. *NeuroImage, 14*, 1013–1026.

Qin, Y., Sohn, M. H., Anderson, J. R., Stenger, V. A., Fissell, K., Goode, A., et al. (2003). *Predicting the Practice Effects on the Blood Oxygenation Level-dependent (BOLD) Function of fMRI in*

a Symbolic Manipulation Task. Proceedings of the National Academy of Sciences of the U.S.A (Vol. 100, pp. 4951–4956).
Schoenfeld, A. H. (1985). *Mathematical problem solving* (pp. 56). Academic Press.
Simon, H. A., & Kaplan, C. A. (1989). *Foundations of cognitive science* (pp. 1–47). MIT Press.
Simon, H. A. (1986). The information processing explanation of gestalt phenomena. *Computers in Human Behavior, 2*, 241–255.
Sohn, M. H., Goode, A., Stenger, V. A., Carter, C. S., & Anderson, J. R. (2003). Competition and representation during memory retrieval: Roles of the prefrontal cortex and the posterior parietal cortex. *Proceedings of National Academy of Sciences, 100*, 7412–7417.
Sohn, M. H., Goode, A., Stenger, V. A., Jung, K. J., Carter, C. S., & Anderson, J. R. (2005). An information processing model of three cortical regions: Evidence in episodic memory retrieval. *NeuroImage, 25*, 21–33.
Wang, J., & Wang, A. (1992). *Cognitive psychology* (pp. 307). Peking University Press.
Wei, X. F., Cui, G. Z., Li, L., & Duan, Y. M. (2011). Classroom interaction analysis based on learning process. *E-education Research*, 12.
Wu, W. J. (1984). Basic principles of mechanical theorem proving in elementary geometries. *Journal of Systems Sciences & Mathematical Sciences, 4*, 207–235.
Zhang, J. Z., Yang, L., & Hou, X. R. (1995). Geometry theorem machine proof of the complete method. *Systems Science and Mathematics*, 3.
Zhang, H. C. C., Dong, Q., Jin, Z., et al. (2004). Multiplication by mental number line: An fMRI study (pp.146). *International Journal of Psychology*, 39.

Chapter 6
Design of Mathematical Inquiry Problems in Primary Schools Based on Cognitive Processes

Countries worldwide attach great importance to exploring the role of mathematics learning and teaching. In recent years, the U.S. government has placed particular emphasis on the importance of involving students in real science inquiry (Loucks-Horsley et al., 2000; Schweingruber et al., 2012). In Europe, Australia, Israel, and other countries, classroom teaching places particular emphasis on inquiry. The National Medium and Long-Te*rm Education Reform and Development Plan* (2010–2020) indicates that school learning provided students with an understanding of society, in-depth thinking skills, and hands-on engagement. Inquiry teaching is an instructional method advocated by the new curriculum reform.

The problem is the basis of mathematical innovation, the starting point of mathematical inquiry learning. In particular, inquiry learning emphasizes the importance of questions in learning activities (Yu, 2004). A proper exploration of questions can stimulate students' curiosity, guide them to think actively, cultivate their ability to think and apply logic, and improve the pertinence and effectiveness of teaching activities. *The Compulsory Education Mathematics Curriculum Standards* (2022 edition) also emphasizes that teachers should become organizers, guides, and collaborators in students' learning activities and provide a good environment and conditions for student development. A typical problem is a necessary condition for inquiry learning. The design of inquiry questions fully reflects the teacher's role as the organizer and guide of student learning activities. When designing the teaching process, teachers need to have a clear direction of inquiry. Primary school research has found that front-line teachers expect to carry out inquiry teaching in the classroom. However, there is a lack of analysis of inquiry problems, and teachers cannot grasp effective methods and means to design inquiry problems. Hence, a common phenomenon occurs in which the form of inquiry is more important than content in classroom teaching. How to create inquiry questions is one of the most common concerns of primary and secondary school mathematics teachers, which is the key to effectively carrying out inquiry teaching and an effective way to improve students' creative thinking ability.

X. Wei, *Artificial Intelligence in Mathematics Education*, Smart Computing and Intelligence, https://doi.org/10.1007/978-981-95-1405-2_6

6.1 Research on Problem Design

In mathematics teaching, the training of students' thinking skills should be incorporated into teaching activities. Teachers cannot take for granted that they can think as students and expect students to answer. Teachers should try to reduce the direct interpretation of teaching materials, which can lead to responses such as "I do not know what to do," so students can experience hands-on application. Teachers should create conditions for students to think independently to gain an understanding of knowledge and truly experience the process of generating knowledge (such as concepts). During this process, teachers should be patient and provide appropriate prompts at the right time.

The learner's learning process is the process of problem solving. Dewey outlines the sequence of events as follows: (1) the presentation of the problem; (2) learners clarify the problem or distinguish the essential characteristics of the problem situation; (3) learners form hypotheses that can be used to solve problems; and (4) learners try to verify the hypothesis until the learner finds the answer to the question (Dewey, 1910). An analysis of the sequence of events solved by the problem shows that only the first step is an external event, and the remainder are internal events; the first step in the external event is the question that is presented to the students. Therefore, in the inquiry learning process, the design of inquiry questions is crucial. The fifth chapter analyzed and simulated the problem-solving cognitive process, that is, the internal events, and provides the basis and reference for the design of inquiry questions. The following sections address the inquiry teaching mode and strategy. To explore the type and design of a problem, information technology can be used to support mathematical inquiry learning.

6.1.1 Exploring Teaching Methods and Strategies

With respect to the inquiry teaching model, Poon proposed a framework for the inquiry teaching of primary school teachers and conducted inquiry teaching practices in four primary schools. Conceptual knowledge and procedural knowledge are very important in exploratory activities (Poon et al., 2012). Wang Jingying used text analysis to compare the inquiry teaching modes of four science teachers in China and the United States. She found that American teachers regard the question as the core and allow students to study implicitly and reflect on the problem through questions. Identifying problems, forming assumptions, making plans, exploring experiments, analyzing data, verifying assumptions, interpreting results, and reviewing and reflecting on practical applications are all aspects of problem-centered circular and open systems. Chinese teachers have focused mostly on the goal of exploring the dimensions of knowledge and skills in teaching (Wang, 2010).

Stafylidou and Vosniadou (2004) reported that students often use invalid or even incorrect strategies when exploring the meaning and size of scores. If the teacher tells the student the answer, inquiry learning becomes a simple knowledge transfer. Teachers need to master the inquiry learning process to provide proper guidance when students explore problems through inquiry.

Fernandez and Yoshida (2012) reported that in Japanese elementary school mathematics inquiry teaching, the inquiry question posed by the teachers is one of the four key elements of classical inquiry teaching. Polygamy argued that mathematics teaching should present students with thought-provoking and controversial questions and that students should try their best to discover as many insights as they can under the given conditions.

6.1.2 Type and Design of Inquiry Questions

Researchers such as Chin and Kayalvizhi (2002), Gott and Duggan (2002), and Watson et al. (1999), and have studied the types of students' inquiry questions and training methods of inquiry learning. Owing to the different research situations, the types of problems examined are quite different. Koufetta-Menicou and Scaife (2000) provided training to improve students' inquiry questions, such as group discussions, brainstorming, and creating interesting situations. Chin and Kayalvizhi (2002) calls noninquiry questions low-level questions and inquiry questions high-level questions. Luo (2010) presents three stages of asking questions: generating question awareness, expressing the problem, and expressing the problem in scientific language.

To explore the issue of inquiry design, Ding et al. (2009) conducted student interviews to eliminate the influence of the viewpoints of physics experts and gain insight on the issue and improve the effectiveness of problem design. Li (2013) analyzed the theoretical framework of designing audience response systems in interactive inquiry teaching. Li (2007) describes the design of the teaching response system in terms of the design steps.

With respect to the educational implications of inquiry questions, Lock (1990)found that many of the questions raised by students were poorly researched and not even educationally valuable. Jong's study of Dutch middle school students revealed that students in inquiry learning classes score high on conceptual knowledge and that students in traditional teaching classes score high on procedural knowledge (De Jong et al., 2010).

6.1.3 Information Technology Support for Mathematical Inquiry Learning

With respect to information technology support for mathematical inquiry teaching, Baki et al. (2011) reported that, compared with traditional teaching, the use of dynamic geometry software improved the spatial recognition skills of first-year normal school students majoring in math. Falcade et al. (2007) used tracking tools in dynamic geometry software to help students explore the concept and trajectory of functions. Eysink et al. (2009)compared the effectiveness of different learning methods supported by technical environments and reported that inquiry learning is most effective in deepening conceptual knowledge. Web-based problem-solving training and assessment systems in IMMEX also emphasize the importance of scientific inquiry and describe the inquiry process (Stevens et al., 2004).

6.1.4 Comments

The above literature analysis indicates that the study of mathematical inquiry has attracted the attention of researchers. Inquiry-based teaching in the United States tends to be more content oriented, whereas inquiry-based teaching in China is more oriented toward teaching methods and teaching strategies. At present, the design of inquiry questions tends to reflect the personal experience of teachers, which is subjective and lacks systematic analysis and scientific design.

6.2 Exploring the Basis and Principles of Problem Design

In June 2001, a basic education curriculum reform in China was announced. The reform introduced changes in the curriculum to emphasize receptive learning, rote learning, and mechanical training status; promote the active participation of students willing to explore and practice; and cultivate the ability to acquire new knowledge to collect and process information, analyze and solve problems and exchange and cooperate. The outline of the national medium- and long-term educational reform and development plan (2010–2020) noted that "the school left students with the practice of understanding society, thinking deeply and practising hands-on." The problem that is used to support the student's cognitive understanding, problem representation and cognitive level should be made consistent to the extent possible with the use of graphical and tabular forms, thus establishing specific procedures and abstract concepts to help students understand the meaning of the problem.

6.2.1 Design Basis of the Inquiry Problem

The design process of primary school mathematics inquiry problems should consider not only primary school children's psychological characteristics and life experiences but also their previous knowledge and problem-solving learning process analysis.

The basic characteristic of primary school children's thinking is that its main form transitions gradually from concrete image thinking to abstract logical thinking. To a great extent, this type of abstract logical thinking is still directly related to sensual experience and still presents as a specific image (Zhu, 2009). Piaget (1952) suggested that pupils are in the transitional stage of the development of concrete thinking to abstract thinking. However, in the information age, in which children are considered "digital natives" (digital natives), there are individual differences in the cognitive development of each child. However, the stage of cognitive development does not change. Throughout primary school, intuitive teaching is an important condition for drawing children's attention. The use of "finger counting" in the process of solving the "counting" problem highlights the role of "object perception" in solving problems among primary school children.

An experimental study revealed that the memory ability of 7- and 8-year-old children is not very different from that of preschool children (Zhu, 2009). With the preliminary development of the unconscious mind and logical memory, the unconscious mind and specific images still occupy a major position. As children enter primary school, awareness, memorization, and abstract logic gradually become dominant cognitive functions.

The cognitive process analysis is based on the mathematical problem-solving cognitive model for primary school. It analyzes and clarifies the cognitive process, providing a clear description. The results of the analysis constitute an important basis for the design of the problem.

6.2.2 Principles of Designing the Inquiry Problem

1. The expression of the problem should conform to the cognitive level of primary school students

An analysis of mathematics textbooks from Grades 1 to 6 revealed that the primary school mathematics curriculum content ranges from concrete to abstract. With increasing grade levels, the degree of abstraction of the course content increases, and the content of the math textbooks in Grades 1 and 2 is composed of specific materials. The design process of inquiry questions should conform to the principles of physical and mental development and the cognitive laws of pupils, reflecting the characteristics of pupils' mathematics learning in primary school. The problem formulation should be consistent with the cognitive level of students and avoid, to the extent possible, an "adult" and "academic" presentation of graphics and forms to help students understand the meaning of the problem and establish specific insights and abstract concepts.

2. The relationship between the problem situation and the real lives of students

Considering the characteristics of pupils' cognitive development, we must teach knowledge in a way that considers students' real-life problems. Students in primary schools, especially those in lower grades, do not understand abstract concepts and do not grasp the regularity of operation rules well. Even in the high school years, we should consider the physical background to which mathematics concepts and rules are attached to enable students to experience and understand them rather than acquiring them through rote learning. Cheung (2008) noted that the inquiry problems that are posed are not related to students' lives, so teachers encounter challenges in finding appropriate research materials. Therefore, the inquiry problem should be designed through an organic combination of knowledge and students' real lives so that it is relatable. Teachers can obtain teaching material from real life to inspire and guide students to enable them to gradually realize that the concepts and laws of mathematics are abstracted from real life. The closer to real life a problem is, the greater the ability of students to understand and apply knowledge and thus achieve more comprehensive, holistic and integrated knowledge, and students' basic knowledge base will also be more robust (Liu, 2002). The knowledge that the students construct in the process of solving practical problems will be flexible. Through solving practical problems, we can help students to experience the value and significance of learning to stimulate their learning motivation and the importance of solving problems in real life (Schliemann, 1985; Schliemann & Nunes, 1990). They reported that solving problems in the context of the real world by constructing a strategy and problem-solving method was more meaningful.

3. Implying knowledge in the process of solving problems

At present, most inquiry questions are often tied directly to the knowledge that students need to learn, but the mere teaching and application of knowledge do not involve "problem inquiry" and "knowledge discovery." Therefore, the knowledge gained is difficult to transfer flexibly into complex practical problems. Students solve inquiry problems to reflect on and abstract professional knowledge and problem-solving strategies. The design of inquiry problems should allow students to "unwittingly" acquire knowledge in the process of solving the inquiry problem.

4. Let the students experience the process of knowledge production

Students need to personally collect data, analyze data, and then discover knowledge to solve a problem. A problem is solved not by "knowing the answer at once" but rather by "thinking hard." Resnick (1987) designates hard and nonalgorithmic thinking an important component of high-level thinking skills. By personally experiencing the knowledge application, students learn how knowledge is used. Knowledge acquired in this way belongs to the students themselves and can be transferred flexibly.

5. Integrating the model mind into the process of problem design

A mathematical model is a common method for solving practical problems in mathematics. Pupils gradually master mathematical models of addition, subtraction, multiplication, division, and equations in mathematics learning. When solving practical problems in real life, we can abstract practical problems into mathematical problems and use mathematical models to solve them. Model thinking provides a basic way for students to understand the relationship between mathematics and the external world, and it is a bridge between basic mathematics knowledge and the application of mathematics. The process of establishing and solving the model should abstract mathematical problems from real life or specific situations. By solving typical problems with model thinking, we can help students initially develop model thinking and improve their interest and application awareness in mathematics learning. When designing a problem, a teacher should fully consider the idea of the model. When students face similar problems in the future, they can abstract them into mathematical problems and use mathematical models to solve them effectively.

6.3 Typical Inquiry Problem Design

Anderson, a modern cognitive psychologist, divides the book *Knowledge of Students' Learning* into "declarative knowledge" and "procedural knowledge" from the perspective of the psychological nature of knowledge. This division has philosophical roots, and Andersen elaborates on his psychological mechanism. The notions of "declarative knowledge" and "procedural knowledge" are common and widely accepted categories of knowledge, and this classification of knowledge is thus also used in this book.

An analysis of the content of the teaching materials revealed that the division between "declarative knowledge" and "procedural knowledge" is not absolute. Research by Rittle-Johnson, Siegler, and Alibali also revealed that the development of declarative knowledge and procedural knowledge are entangled and that there are complex mutual promotions (Rittle-Johnson et al., 2011). According to the different priorities in determining which type belongs, knowledge points can often include both "declarative knowledge" and "procedural knowledge." On the basis of the above analysis, two typical knowledge points, "mode" and "cylindrical flank area," are selected from the mathematics textbooks of Grades 5 and 6. The "mode" in Grade 5 is mainly "declarative knowledge," whereas the "flank area of a cylinder" in Grade 6 is mainly "procedural knowledge."

In the following, the "mode" and "cylindrical flank area" knowledge points are used as examples to discuss the problem design.

6.3.1 "Mode" Precourse Inquiry Question Design

The "mode" is an important concept taught in fifth grade in primary school. Students have previously learned the concepts of "average" and "median."

(一) Theoretical basis

Conceptual learning is not merely the sum of a connection formed by memory or a psychological habit. It is a complex and real thinking activity. Practical experience shows that the direct teaching of concepts is not effective. Although students can remember to explain concepts and imitate conceptual knowledge, their ability to apply concepts to solve real problems is poor. Through experiments, Ach showed that concept formation is not a passive mechanical process but a creative process; a concept is generated and formed during a complex operation, and the purpose of this complex operation is to solve a problem. The external conditions that suggest the mechanical connection of words and objects are insufficient to produce a concept. According to Ach's schema, concept formation is a goal-oriented process. It is a series of operations that serve the various steps leading to the ultimate goal. To ensure that the process of concept formation can proceed, an unsolvable problem must be presented and cannot be solved unless new concepts are formed (Vygotsky, 2010).

Designing and presenting a problem that leads to the formation of a concept does not mean that the problem should be viewed as the reason for the concept formation process. Notably, the goal must be to understand the intrinsic linkages between external tasks and developmental motivation and the formation of concepts as a function of social development and cultural growth. This framing affects not only the content of young people's thinking but also their way of thinking.

The concept of the "mode" is based on data analysis. *The Compulsory Education Mathematics Curriculum Standards* (2011 edition) emphasizes that in mathematics courses, students should give attention to learning the concept of data analysis (Ministry of Education of the People's Republic of China, 2012). Additionally, the 2022 edition stresses that the mathematics curriculum must deliver the appropriate level of instruction, aiming to ignite the interest of students who face challenges in learning, foster their proactive thinking, nurture beneficial study habits, and meet the evolving developmental requirements of the students (Ministry of Education of the People's Republic of China, 2022). Understanding many problems in real life involves performing research, collecting data, making judgments through analysis and interpreting the information contained in the data.

The "mode" is a concept taught in Grade 5. It is directly explained to students. Such teaching is simple and easy, and students can remember the content. However, this teaching method ignores the motivation and development of the concept formation process and ignores the students' experiences in this process. Therefore, in this study, we did not tell the students directly. Instead, we presented the well-designed questions to the students and let them explore the problem-solving process to completion, that is, the process of forming the concept of the "mode" through problem solving.

(二) Design process

1. Textbook title analysis

To teach the concept of "mode" through the teaching materials, a topic is presented before the concept of "mode" is introduced. The purpose of this approach is to enable students to learn the concept of "mode" by solving questions, as shown in Fig. 6.1 (Course Textbook Institute, 2006).

Ten students from Class 2 in Grade 5 were selected to participate in group dance competitions.

The heights of the 20 candidates (unit: m) are as follows:

Fig. 6.1 "Mode" problem in teaching materials

1.32 1.33 1.44 1.45 1.46 1.46 1.47
1.47 1.48 1.48 1.49 1.50 1.51 1.52
1.52 1.52 1.52 1.52 1.52 1.52

According to the above data, which height do you think is most appropriate?

Then, the following information is presented: "In this set of data, 1.52 occurs most frequently. To analyze this set of data, the mode is used. The mode reflects the concentration of values in a set of data."

Analysis of the data in the title and the goal yields the following results:

(1) There is no suitable height;
(2) A single height is not sufficient for choosing 10 students;
(3) The data are given directly, thus eliminating the step of data collection. The actual problem is that the data will not be automatically provided.
(4) The topic of "choosing a dance partner" indicated in the title is not familiar to pupils. It may be familiar to urban pupils, but for rural pupils, this problem situation is highly unusual. It is therefore not conducive to the discovery and mastery of knowledge.

Through the above analysis, we find that the topics presented in the textbooks are not typical examples of the "mode" problem.

2. "Mode" problem-solving cognitive process analysis

The "mode" is a typical part of the "statistics and probability" section and is a concept presented in the second semester of fifth grade in PEP textbooks. Zuyin Ding conducted an experimental study on the process of mastering children's concepts. The results revealed that the concept mastery of primary school children presented stage features. Children in lower grades of primary school tend to use "concrete examples" and "intuitive features" to grasp these concepts. In contrast, children in higher grades of primary school gradually grasp concepts according to nonintuitive "important attributes," "practical functions," and "genus relations." The concept of "mode" trains students to identify nonintuitive "important attributes" of the data, in this case, the most frequent occurrence.

The "mode" is an abstract concept. Although Grade 5 children have some abstract thinking skills, specific materials are still needed to help children understand these concepts. Psychological research also holds that the process of children's mastery of the conceptual system is the process by which children apply various conceptual materials they learned to assimilate (or comprehend) profound and systematic knowledge (Zhu, 2009).

The "mode" cognitive problem-solving process is based on a cognitive model of mathematical problem solving (CMMPS) (Wei & Cui, 2012). Solving for the "mode" in data $\{a_1, a_2, a_3, \ldots, a_n\}$ can be described as follows:

(1) Understanding the question. Find the "mode" in the given data.
(2) Developing a program. Find the most frequently occurring number in the data $\{a_1, a_2, a_3, \ldots, a_n\}$.
(3) Implementing the program.

① Activating the operation of counting in long-term procedural memory. Count the number of $a_1, a_2, a_3, \ldots, a_n$ in data $\{a_1, a_2, a_3, \ldots, a_n\}$.
② Determining that the occurrence times of $a_1, a_2, a_3, \ldots, a_n$ as $M_1, M_2, M_3, \ldots, M_n$, respectively.
③ Activating the comparison operation in long-term program memory, comparing the sizes of $M_1, M_2, M_3, \ldots, M_n$ to determine the maximum value M_{max}.
④ The number corresponding to M_{max} is the mode of the data $\{a_1, a_2, a_3, \ldots, a_n\}$.

(4) Review and check. Check for any errors in each step. The concept of "mode" is added to long-term declarative memory, and the operation for finding the "mode" is added to long-term procedural memory, while the understanding of counting numbers and comparisons are further strengthened.

By analyzing the process of solving the "mode" problem, the key is to determine the problem-solving strategy, that is, "the most frequent number in the data." Performing operations, such as "counting numbers," "comparison," and "corresponding," consists of drawing on previously learned knowledge.

Children's acquisition of concepts is a concrete and gradual abstract process. When children begin to grasp a concept, many concepts are isolated and have not been added to a certain conceptual system due to a lack of knowledge and experience. Only in the conceptual system can the effect improve. For example, within such a system, the concept of "plural" can be mastered more effectively, and the connection between the concepts of median and average can be established only if children have mastered these concepts.

3. Model of the "mode" problem concept

The newly promulgated *Compulsory Education Mathematics Curriculum Standards* (2022 edition) proposed that the mathematics curriculum should emphasize the development of students' model thinking. The process of establishing and solving the model includes the abstraction of mathematical problems from real-life or concrete situations (Ministry of Education of the People's Republic of China, 2022). Reported that asking students to construct a conceptual model of problem solving is a crucial step in problem transformation. Through the analysis of the "mode" solution process, a conceptual model of the "mode" problem is obtained, as shown in Fig. 6.2. Through the analysis of the model, the key to solving the "mode" problem is to choose a strategy for problem solving, that is, how to convert a "birthday" situation into a mathematical problem. After converting to a math problem, students can use their existing knowledge and skills to solve the problem. Therefore, the design of inquiry questions should help students convert application questions into mathematical problems.

4. Inquiry problem of "mode"

In his treatise on "lectures" in *How We Think*, Dewey states, "Preparation is to ask questions and inspire students to think of familiar personal experiences, which is helpful in understanding new issues. As soon as a student associates these insights

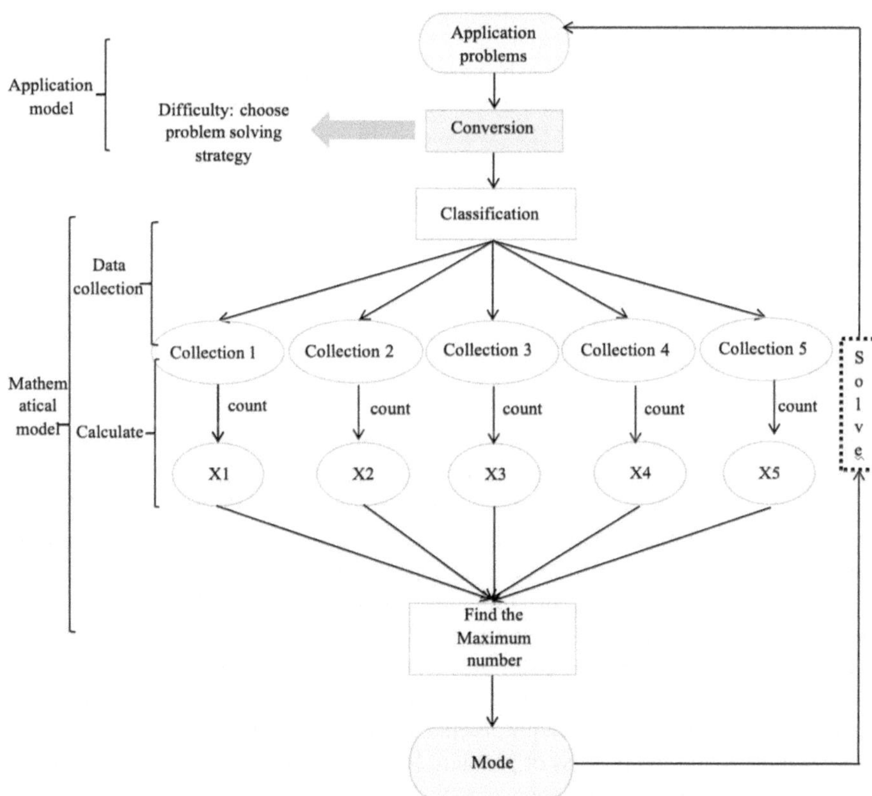

Fig. 6.2 Conceptual model of "Mode"

with actual activities, the process of recognizing something new becomes easier." This statement is applicable to the above considerations and the analysis of the model of solving the "mode" problem. To design the inquiry problem of "mode," the following problem uses a situation familiar to students, such as "birthday."

The school agrees that five (one) classes can have a birthday celebration next year. However, only the birthdays of students born in a certain month can be celebrated. Imagine you are the head teacher:

① How do you choose the month?
② Which month do you think should be chosen?

(1) The concept of design

① This topic is a real application problem, and the "birthday" situation is closely related to students' actual lives.
② Students analyze problems, build a model of the problem, and choose problem-solving strategies independently;
③ Students collect data and obtain the number of people who have a birthday each month independently;
④ Students calculate statistics and the number of people with a birthday each month independently;
⑤ When the number of students who have birthdays in the same month is compared, the month with the most birthdays is the solution to the problem.

(2) The goal of design

① Enable students to not only understand the concept itself but also experience the meaning of the concept.
② Enable students to experience the "mode" concept through the application of the problem situation. Moreover, by building several strategies and methods for problem solving, knowledge transfer can be deepened.
③ Train students on converting the application model into a mathematical modeling problem.

By solving the "mode" inquiry question, teachers do not convey the concept of the majority directly; rather, they allow students to calculate their own statistics for students' birthday months in the class. Moreover, the number of birthdays for each month is compared. Students not only understand the concept itself but also experience the meaning of the concept; experience the application of the concept of "mode" by building relevant strategies and methods, thereby deepening their knowledge transfer; and learn how to convert the application model into a mathematical modeling problem. The design of the "mode" inquiry questions embodies the "problem context–model building–solution verification" mathematical activities process and reflects the basic requirements of the model. It is conducive to helping students understand the process of problem solving, grasp knowledge of the "mode" concept and accumulate experience in math activities. Solving the inquiry problem is conducive to helping students take the initiative to find, analyze and solve problems to train their innovative awareness.

6.3.2 *"Cylinder Flank Area" Before the Inquiry Questions are Designed*

(一) Theoretical basis

Procedural knowledge refers to behavior performed under certain conditions and usually refers to the ability to operate. The "cylindrical flank area" is a typical form of procedural knowledge. Procedural knowledge is obtained through practice. In many theories about problem solving and skill acquisition, declarative knowledge is often used as procedural knowledge of preparatory knowledge and existence (Anderson, 1983; Byrnes, 1992).

When learners encounter new problem situations, the process of problem solving is the process of obtaining advanced rules. Gagne's research on problem solving shows that when a learner succeeds in solving a problem, a high-level rule is obtained. This rule can be quickly generalized to similar problems. The direct presentation of the answer to the learner is ineffective for learning. The reason is that such a presentation does not require the acquisition of advanced rules, and the answer can be effectively learned as a simple chain. The most reliable teaching method is to use examples to stimulate learners to discover rules by themselves (Gagné, 1999). High-quality pretest questions are designed, allowing students to explore and experience the process of knowledge discovery, that is, the process of obtaining high-level rules.

The "flank area of the cylinder" is a new knowledge point taught in sixth grade. Before students learn about the area of a rectangle, the circumference of a circle, the degree of a circle, and related concepts, they acquire the prerequisite knowledge needed to solve the "cylinder flank area." Thus, this study designed typical problems and allowed the students to determine how to calculate the flank area of the cylinder.

(二) Design process

The "cylinder side area" is procedural knowledge taught in the sixth grade of elementary school. Before students learned "the circumference of a circle" and "the area of a rectangle," other calculations are performed.

1. Textbook title analysis

Teaching materials for the "cylinder flank area" (as shown in Figs. 6.3 and 6.4) allows students to learn the cylinder, given the "bottom" "flank" "height" and other concepts. Moreover, after expanding the cylindrical side to be a rectangle, the calculation of the flank area of the cylinder is converted into that of the area of a rectangle.

This design of teaching materials is logically in line with students' cognitive rules. A careful analysis of the details reveals the following:

(1) The strategy of unfolding the side of a cylinder is presented directly to the student, and the student does not experience in-depth thinking on "why."
(2) The title should present a situation and allow students to develop a method for calculating the flank area of the cylinder.

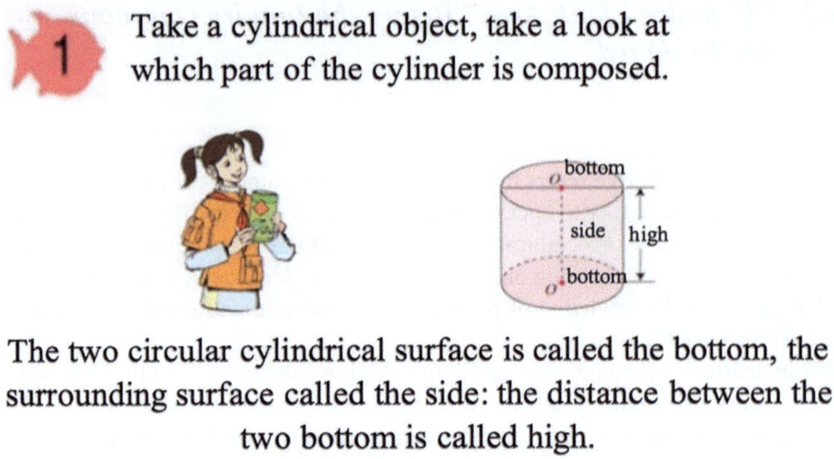

Fig. 6.3 "Cylinder flank area" Problem 1 in the textbook

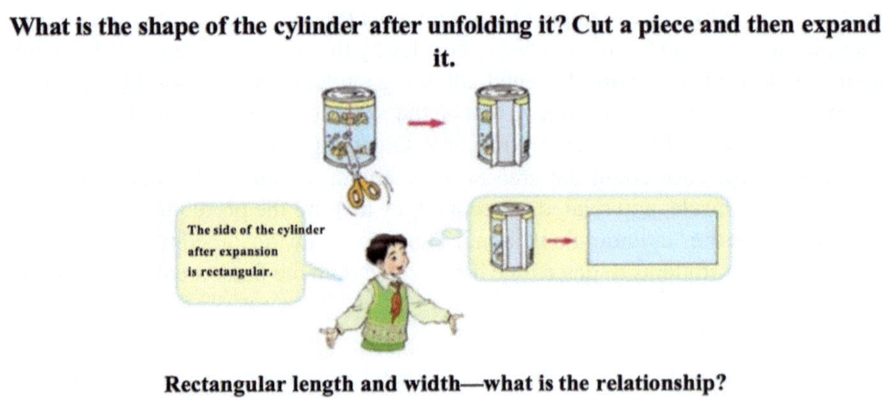

Fig. 6.4 "Cylinder flank area" Problem 2

2. Problem-solving model

The "cylindrical side area" problem-solving model is shown in Fig. 6.5.

An analysis of the "cylinder flank area" reveals that the key to solving the problem of the cylinder side area is converting it to a rectangular area. Designing problems and creating situations from the perspective of students' lives so that students can generate an idea for converting a cylinder flank area into a rectangular area in the process of solving the problem.

3. Inquiry questions

On the basis of the inquiry problem design principles and the "cylinder flank area" problem-solving model, a series of questions were posed: A potato chip factory

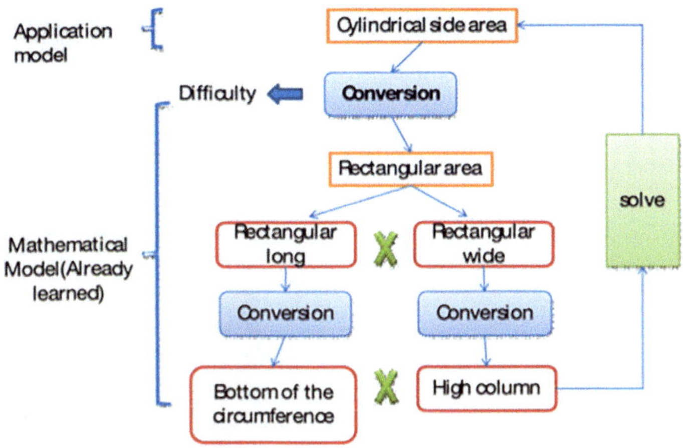

Figure 6-5 "Cylindrical side area problem" solving model

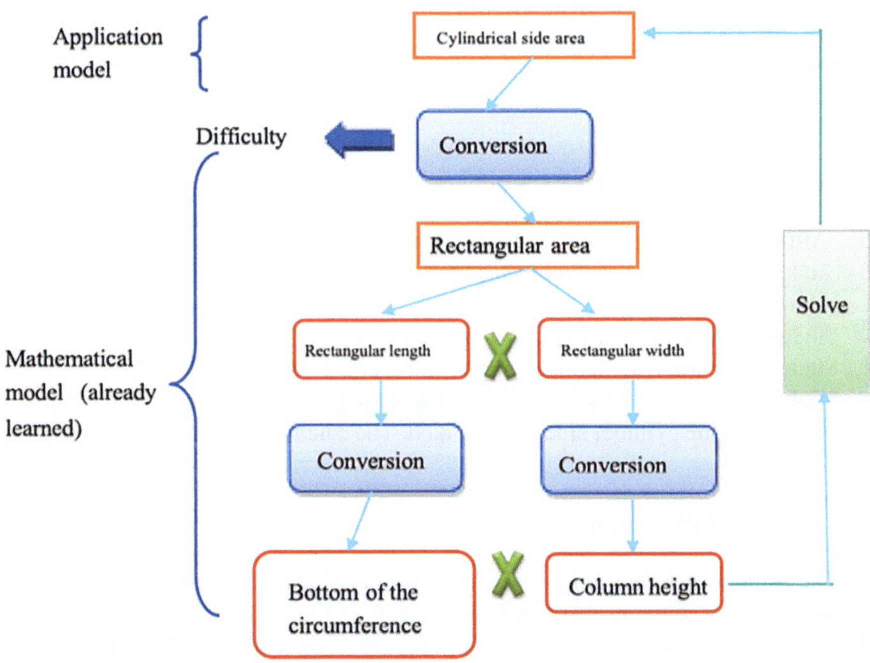

Fig. 6.5 Model for solving the "Cylindrical side area" problem

Fig. 6.6 Chip box

produces a batch of potato chips that need to be packaged on the side of the potato chip box (as shown in Fig. 6.6). Can you help the factory calculate how many wrappers are needed for each box?

(1) The concept of design

① "Affixing wrapping paper" is closely related to the actual life situation of students.
② Students analyze the problem, build a model of the problem, and select a problem-solving strategy.
③ Students cut paper and pack chip boxes.
④ Students measure the amount of wrapping paper required.
⑤ Students calculate the area of the wrapping paper, that is, length × width;
⑥ The area of the cylinder side is calculated. The calculation method is summarized below.

(2) The goal of the design

① The students not only understand the cylinder side area calculation formula but also experience the process and significance of the formula.
② The students experience the application of the formula "cylinder side area" and determine the "side area of the cylinder" independently to solve the problem, building strategies and methods to deepen their knowledge of the relevant mathematical concepts.

6.4 The Teaching Application of the Inquiry Problem

Training students' logical reasoning ability is advocated by *The Compulsory Education Mathematics Curriculum Standards* (2022 edition) and is also one of the goals of primary mathematics inquiry teaching. The training method uses typical problems in classroom inquiry teaching so that students can improve their logical reasoning skills in solving typical inquiry questions.

To validate the teaching effectiveness of inquiry questions, on the basis of the method of the inquiry problem design method described in the book, typical inquiry questions are designed for all knowledge points covered in fourth-grade mathematics and used in inquiry teaching in the mathematics classroom. For this study, we selected two fourth-grade primary schools in Beijing, one for the experimental class and one for the comparison class, as shown in Table 6.1. We used typical inquiry questions in the experimental class. There were 34 students in the experimental class, including 21 boys and 13 girls. Their ages ranged from 10–11 years; 32 students were in the contrast class, including 17 boys and 15 girls, and their ages were between 10 and 11 years.

During the experiment, the Raven Standard Progressive Matrices Test was used to test the students' logical reasoning ability. The pretest was conducted in January 2013, and the posttest was conducted in July 2013. The experiment lasted one semester. The earliest Raven Standard Progressive Matrices Test scale was the nontext intelligence test designed by the British psychologist Raven (J.C. Raven) in 1938. The scale consists of a total of five modules, A–E, of 60 pictures. The difficulty of the modules increase; thus, the A module is the simplest, and the E module is the most difficult. Pre- and posttests were used to test the students' progress, and SPSS 19 was used for data analysis.

6.4.1 Comparison of Reasoning Ability Between the Experimental Class and Comparison Class

The pretest and posttest data of the Raven scale are shown in Table 6.2.

As shown in Table 6.2, the means of the Raven scores in the pretest, experimental and control classes are very close (the difference is only 0.8), whereas mean for the experimental group in the posttest phase is 5.4 higher than that of the comparison class.

Table 6.1 Student situation statistics (Unit: person)

	Boys	Girls	Age (10–11 years old)	Total
Experimental class	21	13	34	34
Contrast class	17	15	32	32

Table 6.2 Pretest and posttest Raven scores

		Mean value	Standard deviation	Standard error	Minimum value	Maximum
Pretest Raven scores	Comparison class	40.05	7.93	1.82	23	53
	Experimental class	40.85	6.54	1.12	31	55
Posttest Raven scores	Comparison class	42.84	6.64	1.52	24	52
	Experimental class	48.24	5.03	0.86	36	58

Table 6.3 Variance analysis

	Quadratic sum	df	Mean square	F	Significance
Pretest Raven scores	7.807	1	7.807	0.16	0.69
Posttest Raven scores	354.526	1	354.526	11.09	0.00

The typical inquiry question in the teaching process was used as the independent variable, and factor analysis of single variance was used to compare the pretest and posttest Raven scores in the experimental class and the comparative class. The results are shown in Table 6.3.

Table 6.3 shows that in the pretest stage, $F = 0.156$, the significance level is $p = 0.69 > 0.05$, and there is no significant difference between the experimental class and the comparison class. In the posttest phase, $F = 11.09$, the significance level is $p = 0.00 < 0.05$, and the Raven results of the experimental class and comparison class were significantly different. These findings indicate that the use of typical inquiry questions effectively improves the inference ability of experimental students.

6.4.2 Comparative Analysis of Mathematical Reasoning Ability in the Experimental Class

Considering that age has an effect on grades, the score in the sample was compared with the norm for age to exclude the impact of aging. Consequently, the score could be categorized into eight levels, namely, the Raven grade. The pretest and posttest Raven scores and Raven grades of the experimental class are shown in Table 6.4.

Paired samples t tests were performed on the pretest and posttest data of the experimental class, as shown in Table 6.5.

The data in Tables 6.4 and 6.5 reveal that the experimental class reasoning ability paired t test sample number was 34, the average pretest Raven score was 40.85, the

Table 6.4 Experimental class Raven score and Raven grade pretest and posttest means

	Mean value	N	Standard deviation	Standard error of mean value
Pretest Raven scores	40.85	34	6.54	1.12
Posttest Raven scores	48.24	34	5.03	0.86
Pretest Raven scores	4.29	34	1.19	0.21
Posttest Raven scores	5.41	34	1.26	0.22

Table 6.5 T tests of paired samples in the experimental class

	Paired difference			t	df	Sig (Bilateral)
	Mean value	Standard deviation	Standard error of mean value			
Pretest Raven scores—Posttest Raven scores	−7.39	4.78	0.82	−9.01	33	0.00
Pretest Raven scores—Posttest Raven scores	−1.12	2.20	0.21	−5.43	33	0.00
Pretest A module—Posttest A module	−0.38	0.74	0.13	−3.02	33	0.01
Pretest B module—Posttest B module	−0.71	1.51	0.26	−2.73	33	0.01
Pretest C module—Posttest C module	−1.32	1.61	0.28	−4.80	33	0.00
Pretest D module—Posttest D module	−1.74	2.09	0.36	−4.83	33	0.00
Pretest E module—Posttest E module	−3.24	2.16	0.37	−8.73	33	0.00

average posttest Raven score was 48.24, and the average Raven score increased by 7.39 from pretest to posttest. The significance level $p = 0.00 < 0.001$ reached a very significant level. In other words, when typical inquiry questions were used to teach, students significantly improved their mathematical reasoning ability. The standard deviation of the posttest data was less than the standard deviation of the pretest data; that is, the degree of dispersion decreased, and the average value increased, indicating that the scores of the low-scoring students in the pretest improved. In other words, using typical inquiry questions improves the learning performance of students with poor mathematical reasoning ability.

An analysis of the Raven grade pretest and posttest data revealed that the average pretest Raven grade was 4.29, and the average posttest Raven grade was 5.41. In other words, the average levels of the pretest and posttest middle school students' Raven grades were in the upper middle range. The average posttest level was 1.12 higher than the average pretest level, which is a difference higher than one level, and the significance level $p = 0.00 < 0.001$ indicated extreme significance. In other words, the use of typical inquiry questions before and after teaching significantly improved students' mathematical reasoning ability.

Further analysis of the pretest and posttest data of the five modules of A-E in the Raven scale revealed that the promotion of the three modules of C, D, and E was more significant, with a significance level of $p = 0.00 < 0.001$, indicating a very significant level. Part E, the most difficult part, showed the most significant improvement. These findings indicate that the use of typical inquiry questions in primary school mathematics classroom teaching plays an important role in improving students' advanced mathematical reasoning ability.

6.5 Summary

Mathematics teaching advocates inquiry learning, and the inquiry problem is the key to effectively implementing inquiry learning. How to design a typical inquiry problem is a common concern for the majority of primary and secondary school mathematics teachers and is an important part of inquiry teaching. Teachers carry out effective interactions in the process of student inquiry and conduct targeted guidance according to the characteristics of the inquiry problem. In the process of inquiry teaching, the design of typical inquiry questions needs to fully promote students' enthusiasm, relate to their real lives, and correspond to students' physical and mental development and the characteristics of the mathematics curriculum.

This chapter discusses research on inquiry problem design from the aspects of the inquiry teaching mode and strategy, the type and design of inquiry questions, and mathematical inquiry learning supported by information technology. The basis and principles of primary school mathematics inquiry question design are proposed as follows: (1) The problem statement is in line with the cognitive level of primary school students. (2) The problem situation is related to students' real lives. (3) Knowledge is implied in the process of solving problems. (4) Students experience the process of knowledge generation. (5) The model idea is integrated into the problem design process. This chapter is based on these design principles, given the preclass exploration question design process of the "mode" and "cylindrical side area." Accordingly, the typical inquiry questions are designed for all knowledge points of fourth-grade mathematics, an experimental teaching method is explored in the mathematics classroom, and the experimental results are analyzed.

The reasonable use of inquiry questions in the elementary school mathematics classroom arouses students' interest in mathematics learning, attaches importance to students' experience in exploring mathematics, promotes and improves students' mathematical thinking ability, encourages students' creative thinking ability, proactively cultivates good mathematical study habits, improves their ability to use mathematical knowledge to solve practical problems, provides a basis and reference for the application of technology in mathematics inquiry teaching, and provides a basis for conducting "one-on-one" cognitive diagnosis.

References

Anderson, J. P. (1983). *The architecture of cognition* (pp. 59). Harvard University Press.

Baki, A., Kosa, T., & Guven, B. (2011). A comparative study of the effects of using dynamic geometry software and physical manipulatives on the spatial visualization skills of preservice mathematics teachers. *British Journal of Educational Technology, 42*, 291–310.

Byrnes, J. P. (1992). The conception basis of procedural learning. *Cognitive Development, 7*, 235–257.

Cheung, D. (2008). Facilitating chemistry teachers to implement inquiry-based laboratory work. *International Journal of Science and Mathematics Education, 6*, 107–130.

Chin, C., & Kayalvizhi, G. (2002). Posing problems for open investigations: What questions do pupils ask? *Research in Science & Technological Education, 20*, 269–287.

Course Textbook Institute. (2006). *Compulsory education curriculum standard experimental textbook mathematics* (grade five). People's Education Press.

De Jong, T., Hendrikse, P., & van der Meij, H. (2010). Learning mathematics through inquiry: A large-scale evaluation. In M. Jacobson & P. Reimann (Eds.), *Designs for learning environments of the future*. Springer.

Dewey, J. (1910). *How we think* (pp. 27). Health.

Ding, L., Reay, N. W., Lee, A., et al. (2009). Are we asking the right questions? Validating clicker question sequences by student interviews. *American Journal of Physics, 77*, 643.

Eysink, T. H. S., De Jong, T., Berthold, K., et al. (2009). Learner performance in multimedia learning arrangements: An analysis across instructional approaches. *American Educational Research Journal, 46*, 1107–1149.

Falcade, R., Laborde, C., & Mariotti, M. A. (2007). Approaching functions: Cabritools as instruments of semiotic mediation. *Educational Studies in Mathematics, 66*, 317–333.

Fernandez, C., & Yoshida, M. (2012). *Lesson study: A Japanese approach to improving mathematics teaching and learning* (pp. 112). Routledge Press.

Gagné, R. M. (1999). *Learning conditions and teaching theory* (Pi, L. S., Wang, Y. X., & Zhu, K., et al., Trans.). East China Normal University Press.

Georgia, G. (1987). *The discovery of mathematics* (Vol. II) (Y. T. Liu, & Q. Zhang, Trans.). Science Press.

Gott, R., & Duggan, S. (2002). Problems with the assessment of performance in practical science: Which way now? *Cambridge Journal of Education, 32*, 183–201.

Jonassen, D. H. (2003). Designing research-based instruction for story problems. *Educational Psychology Review, 15*, 267–296.

Koufetta-Menicou, C., & Scaife, J. (2000). Teachers' questions-types and significance in science education. *School Science Review, 81*, 79–84.

Li, H. M. (2013). Teaching answering system problem design framework and strategy. *Electrochemical Education Research*, 8.

Liu, R. D. (2002). The application of problem-based learning in primary and secondary schools. *Journal of East China Normal University (Educational Science Edition)*, 3.

Lock, R. (1990). Open-ended problem living investigations. *School Science Review, 71*, 63–72.

Loucks-Horsley, S., & Olson, S. (Eds.). (2000). *Inquiry and the national science education standards: A guide for teaching and learning* (pp. 26). National Academies Press.

Luo, G. Z. (2010). An empirical study on the issue of inquiry. *Course Teaching Method*, 30.
Li, P. F. (2007). *Creating and evaluating a new clicker methodology* (pp. 102–107). The Ohio State University.
Ministry of Education of the People's Republic of China. (2012). *Compulsory education mathematics curriculum standards* (2011 edition). Beijing Normal University Press.
Ministry of Education of the People's Republic of China. (2022). *The compulsory education mathematics curriculum standards* (2022 Edition). Beijing Normal University Press.
Piaget, J. (1952). *The child's conception of number* (pp. 49). Routledge & Kegan Paul.
Poon, C. L., Lee, Y. J., Tan, A. L., et al. (2012). Knowing inquiry as practice and theory: Developing a pedagogical framework with elementary school teachers. *Research in Science Education, 42*, 303–327.
Resnick, L. B. (1987). *Education and learning to think* (pp. 59). National Academy Press.
Rittle-Johnson, B., Siegler, R. S., & Alibali, M. W. (2011). Developing conceptual understanding and procedural skill in mathematics: An iterative process. *Journal of Educational Psychology, 93*, 346–362.
Schliemann, A. D., & Nunes, T. (1990). A situated schema of proportionality. *British Journal of Developmental Psychology, 8*, 259–268.
Schliemann, A. D. (1985). Mathematics among carpenters and carpenters apprentices: Implications for school teaching. In *Mathematics for all* (pp. 92–95).
Schweingruber, H., Keller, T., & Quinn, H. (Eds.). (2012). *A framework for K-12 science education: Practices, crosscutting concepts, and core ideas*. National Academies Press.
Stafylidou, S., & Vosniadou, S. (2004). The development of students' understanding of the numerical value of fractions. *Learning and Instruction, 14*, 503–518.
Stevens, R., Soller, A., Cooper, M., et al. (2004). Modeling the development of problem solving skills in chemistry with a web-based tutor. In *Intelligent tutoring systems* (pp. 580–591). Berlin.
Vygotsky, L. S. (2010). *Thinking and language* (W. Li, Trans.). Peking University Press.
Wang, J. Y. (2010). Comparison of objectives of inquiry teaching programs between Chinese and American science teachers. *Global Education Prospects*, 4.
Watson, R., Goldsworthy, A., & Wood-Robinson, V. (1999). What is not fair with investigations? *School Science Review, 80*, 101–106.
Wei, X. F., & Cui, G. Z. (2012). Primary school mathematics problem solving cognitive model. *Electrochemical Education Research*, 11.
Yu, W. S. (2004). On autonomy, cooperation, inquiry learning. *Education Research*, 11.
Zhu, Z. X. (2009). *Children's psychology*. People's Education Press.

Part III
Application of Cognitive Simulation

Chapter 7
"One-on-One" Cognitive Diagnosis

Cognitive diagnosis is the application of cognitive analysis in teaching. On the basis of the problem-solving cognitive process analysis and cognitive simulation discussed in Chaps. 4 and 5, we propose the following to deepen the understanding cognitive operations and cognitive components in the process of problem solving and to explore the cognitive process of "one-on-one" cognitive diagnosis on the basis of a cognitive model. Specifically, more detailed and more targeted guidance and advice for educational practice and individual development can be provided to students, especially those with learning difficulties.

Many countries stress the importance of improving student achievement. In January 2001, the Bush Administration passed the No Child Left Behind Act (NCLB) (U. S Department of Education, 2001) in the United States. This act listed "improving the academic performance of underprivileged students" as the first area of seven priorities. It proposed that public education evaluation should provide descriptive and diagnostic reports to each student, parent, and teacher. To better meet the specific requirements of the bill, the Obama administration later approved the Elementary and Secondary Education Act (ESEA) (U.S Department of Education, 1965) in September 2011 to better focus on improving students' learning and teaching quality.

Compulsory education in China also greatly emphasizes the development of all students and highlights the purpose of learning evaluations. For example, *The Compulsory Education Mathematics Curriculum Standards* (2022 edition) (Ministry of Education of the People's Republic of China, 2022) highlights how the mathematics curriculum should provide every student with good math education and allow different people to distinguish mathematical development. The main purpose of a learning evaluation is to fully understand the process and results of student learning, encourage student learning and improve teachers' instructional methods. The past form of study evaluation, which focused on the results and ignored the learning process, has been changed.

An essential element of education is evaluation. Learning evaluations focus on the learning process, and problem solving is an important part of learning. The

X. Wei, *Artificial Intelligence in Mathematics Education*, Smart Computing and Intelligence, https://doi.org/10.1007/978-981-95-1405-2_7

theories and methods of cognitive diagnosis (CD) are inseparable components when measuring and evaluating the cognitive process of students' problem solving. Cognitive diagnosis emphasizes the learning process and identifies deficiencies in the process. The diagnostic results are provided to students, parents, and teachers to help students improve their learning and to improve teachers' teaching quality.

7.1 Cognitive Diagnosis of Primary School Mathematical Problems

Solving mathematical problems has always been the focus of research in psychology and mathematics education in China and abroad (Chen et al., 2004). Montague studied students in Grade 6 with learning difficulties and the role of cognitive and metacognitive strategies and subsequently proposed cognitive and metacognitive teaching models to solve applied problems (Montague, 1992).In this model, Montague broke down the cognitive process of solving problems into seven stages: reading, analyzing, visualizing, presuming, estimating, calculating, and checking. Jitendra applied schema teaching to study children with poor learning performance and conducted systematic research. She used pictorial instruction to develop effective interventions for students with poor performance on application problems in math. With the exception of a few individual cases, most students' problem-solving strategies improved.

Tatsuoka employed diagnostic tests to compare the math scores of eighth graders from 20 countries and analyzed the data via the Third International Math and Science Study-Revised. Tu et al. (2010) applied cognitive diagnostics to study the process by which primary school children solve mathematical problems and explored the characteristics of children's cognitive development with respect to their mathematical problem-solving ability and shortcomings to promote children's cognitive development and knowledge acquisition. These studies provide an important theoretical basis for studying the cognitive diagnosis of primary school mathematics problem solving.

7.1.1 Cognitive Diagnosis Analysis of Students Answering the Application Questions

Many scholars have analyzed errors in problem solving with respect to mathematical problem statements. Hayes et al. reported that if students were able to correctly judge the overall sentence structure of math application problems, then they could correctly identify the necessary information amid redundancy or irrelevance in a long text description. However, students had difficulty identifying necessary information if they misjudged the type of problem, and students could be easily led to make mistakes in problem solving. Meyer's study revealed that, overall, students are more likely to correctly conclude that relevant information is necessary (accuracy rate of 89%) but have difficulty understanding that irrelevant information is unnecessary (accuracy rate of 54%) (Mayer, 1987). He and Fu (1995) analyzed the question representation and problem-solving results of 34 college students and reported the relationship between the problem statement and problem-solving errors. Anand and Ross (1987) reported that the major reason for students' errors in solving application problems was their misinterpretation of the problem statement rather than any calculating difficulties. Shi Tieru collected and analyzed the data collected from oral problem-solving reports and found that a crucial reason for incorrect problem solving is the inability to correctly identify the model, which often caused students to blindly try to solve the problem and repetitively search for solutions (Shi, 1985). Kotovsky et al. (1985) studied the different characteristics of problem isomorphs and systematically analyzed the reasons for the different difficulty levels of the problems. The experimental results showed that the problem structure cannot explain the difficulty of the problem from the perspective of the problem structure. The difference in the ways in which problems are imagined, constructed or considered is the key to determining their difficulty.

The studies above analyzed the relationship between problem presentation and problem-solving errors and identified the causes of problem-solving errors, but they did not provide pertinent measures to help students correct mistakes.

7.1.2 *Mathematical Cognitive Diagnosis Through Cognitive Neuroscience*

Cognitive neuroscience employs brain imaging and other methods to explore the core of mathematical cognition and to make new interpretations of the physiological processes and environments that address the barriers of mathematical problems. According to Rotzer et al., brain regions associated with processing the size of numbers in children with impaired computing ability present structural abnormalities in the brain. De Smedt et al. (2009) reported that human chromosome 22q11 deletion syndrome was associated with maladjustment. Kaufmann et al. (2009) reported that brain regions associated with processing the number sizes in children whose brains had impaired computing ability presented structural abnormalities.

With increasing research on mathematical cognition in cognitive neuroscience, cognitive neuroscience and math education are being effectively combined. Cognitive neuroscience methods are used to provide a reliable basis for the assessment of mathematical cognitive impairment and effective interventions for diagnosing the reasons for mathematical cognitive impairment, thereby helping students with cognitive impairment improve their math learning performance.

Nearly all domestic researchers use the latest or several recent math scores of students as indices to measure students' mathematical ability. However, examinations are usually given periodically; they generally correspond to the knowledge students learned during a certain period of time, which is related to the degree of difficulty of the questions. However, it is unclear whether mathematics scores can reflect students' authentic mathematical ability. The diagnosis of mathematical impairment should be established on the basis of a comprehensive and accurate assessment of students' mathematical ability.

Therefore, when diagnosing mathematical impairment, the student's cognitive process of problem solving should be focused on rather than just the result of problem solving. Through the analysis of the problem-solving cognitive process, we can objectively identify problem-solving difficulties and perform targeted diagnosis and intervention.

7.2 The Purpose and Characteristics of "One-on-One" Cognitive Diagnosis

Most traditional tests are paper and pencil tests. Traditional tests often adopt the form of "one test for all" or "one question for all," and researchers' attention often focuses on the test results. As tests measure only test scores or ability scores, the information is simple. With a single score, teachers and researchers cannot identify the knowledge that students have mastered and that which they have not, and researchers cannot analyze the reasons for students' problem-solving errors. For students with the same test score, it is impossible to determine the differences in knowledge mastery and cognitive processing that may exist between them. The exam is not designed to evaluate students' abilities but rather to identify the problem more accurately.

The purpose of cognitive diagnosis is to diagnose students' advantages and disadvantages, especially their shortcomings and difficulties in solving mathematical problems, and 25 provides a reliable basis for teaching interventions.

Although the results of cognitive diagnosis provide a reliable basis for teaching interventions, they do not provide an implementation plan for how to use these interventions in student learning and teaching. However, difficulties are often encountered when diagnostic findings are used in teaching to motivate students to learn and improve their learning performance. Roussos et al. (2007) researched the validity of cognitive diagnosis through investigations and interviews and reported that teachers and students had a difficult time integrating diagnostic results into learning and teaching. Polia believed that diagnosis is a more detailed assessment of a student's learning. To improve students' academic performance, teachers need to assess students' strengths and weaknesses in detail (Tu et al., 2007). According to Vygotsky's "recent development zone" theory, different children (or the same children) can have different areas of recent development zones in different cognitive domains and benefit from teaching in different contexts. Geert P. V. (1998) advocates that diagnosis and teaching should be based on children's current and future levels of development. DiBello and Stout (2007) emphasizes that the purpose of an exam is to provide students with the information they need to understand the problem and to provide information to be used directly for instructional and student learning. Lane S. (2004), Leighton (2004) reported that few large-scale tests generate diagnostic information on a candidate's thought process. Few large-scale tests have clear inference goals. Therefore, the diagnosis of different levels of student intervention is necessary.

The Compulsory Education Mathematics Curriculum Standards (2011 edition) emphasizes that the main purpose of evaluation is to fully understand the process and results of student learning and to encourage students to learn and teachers to improve teaching. Moreover, the 2022 edition *underscores the importance of guiding students to discover and articulate problems within real-world scenarios.* Evaluations should give attention to the results of student learning and the learning process(Ministry of Education of the People's Republic of China, 2022). "One-on-one" cognitive diagnosis is a dynamic assessment method that combines teaching and diagnosis to meet the different levels of children's current and future development. It is characterized by an emphasis on a "one-on-one" orientation, a balance between the assessment of learning outcomes and the analysis of the learning process, and a combination of identification and classification, diagnosis and prescription. The aim is to assess each student's strengths and weaknesses in an in-depth manner and to provide timely and appropriate feedback on the student's problem-solving performance to guide them through the process, help them achieve their goals and facilitate learning.

The Compulsory Education Mathematics Curriculum Standards (2022 edition) also emphasizes that teaching activities should strive to help all students to meet the basic requirements of the curriculum objectives while focusing on the individual differences of students and promoting the development of each student on an individual basis. For students with learning difficulties, teachers should give timely attention and assistance to encourage them to take the initiative to participate in mathematical learning activities, try to solve the problem on their own, and express their views. Teachers must promptly acknowledge students' progress, patiently guide them to analyze the causes of their difficulties or mistakes, and encourage them to correct their mistakes to increase their interest and confidence in learning mathematics (Ministry of Education of the People's Republic of China, 2022). "One-on-one" cognitive diagnosis focuses extra attention to the learning process of students with learning difficulties and records and analyzes the cognitive changes of students during different stages of problem solving.

7.3 "One-on-One" Diagnosis Based on the Cognitive Model

7.3.1 Diagnostic Process

To study children's cognitive development, Piaget adopted multiple research methods, such as observation and clinical methods. Campbell and Carlson stated that using these methods to study cognitive diagnosis has great significance (Campbell & Carlson, 1995).

Researchers (Kane, 1992; Messick, 1989 have argued that the cognitive model of learning is an important component of describing test structures, designing test questions, and generating diagnostic inferences on the basis of test scores. In the process of "one-on-one" cognitive diagnosis, we design typical problems on the target knowledge points, analyze the correct problem-solving cognitive process and students' actual cognitive process in problem solving on the basis of CMMPS, compare the similarities and differences between the two processes, and use the results to determine students' problems. The cognitive model-based "one-on-one" cognitive diagnosis is shown in Fig. 7.1.

(1) Cognitive process analysis of problem solving via CMMPS

Yang et al. suggested that cognitive diagnosis tests in education or psychology should measure at least three aspects of cognitive characteristics: (1) The more important knowledge or skills in a particular cognitive field, which are the foundation for higher level ability construction; (2) the knowledge structure, which indicates not only the quantity of knowledge and skills but also how people organize these types of knowledge and skills; and (3) the cognitive process Yang & Embretson, (2007). The "one-on-one" cognitive diagnosis fully considers these three aspects of cognitive characteristics: the analysis of the primary school mathematics problem-solving cognitive process is the foundation, the mastery of the knowledge and skills needed to solve elementary school mathematics problems is the key, and the knowledge of elementary school mathematics problem-solving knowledge or skills is the guarantee. These bases of the CMMPS analysis of typical problems can help researchers obtain the correct problem-solving cognitive process.

(2) Data collection via oral reports

Leighton & Gierl, (2007) applied the oral reporting method to collect data to use as the basis of cognitive diagnosis. In "one-on-one" cognitive diagnosis research, oral reports are used to collect data from students during the process of problem solving. Then, on the basis of the CMMPS analysis of the oral report data, the actual problem-solving cognitive process is obtained.

(3) Formulation of a diagnostic scheme

By comparing and analyzing the results of (1) and (2), we identified the cognitive process that led to the incorrect solution of the problem and formulated a diagnostic scheme accordingly.

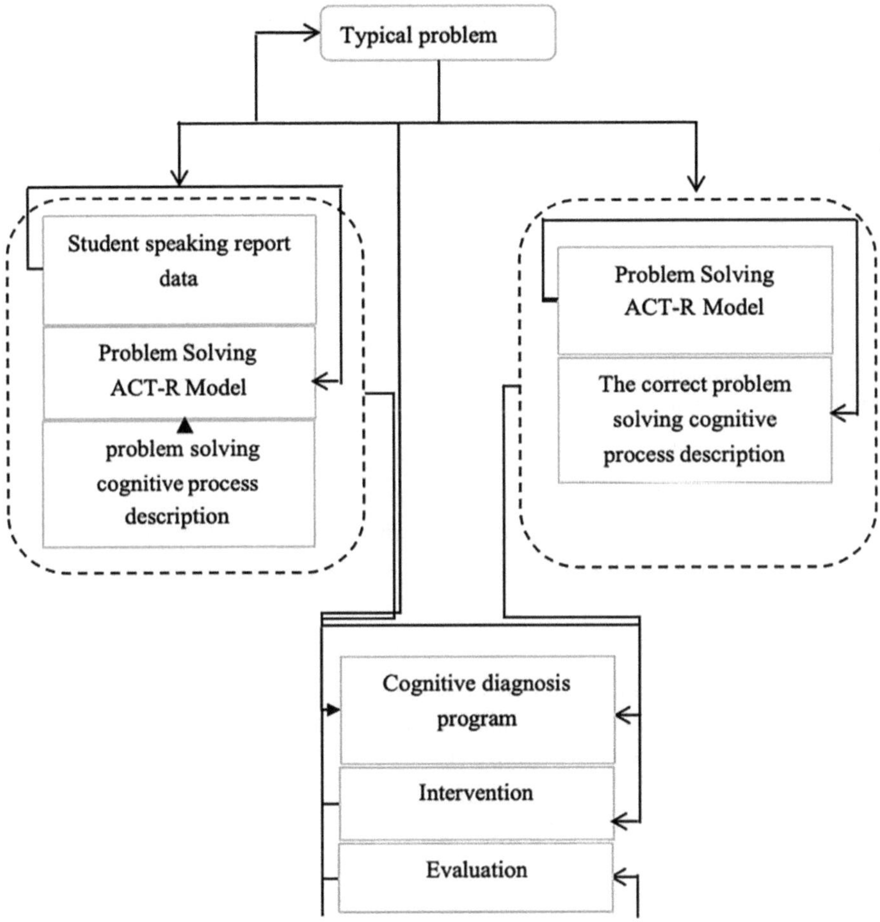

Fig. 7.1 Model-based "one-on-one" cognitive diagnosis

(4) Intervention with students

The intervention process draws on the famous mathematician and mathematics educator Polya's idea of the four stages of mathematical problem solving.

First, understand the problem. It is important to understand the problem statement and clearly know what the problem is asking.

Second, arrange a plan. It is important to understand how each item is related, including the relationship between the unknown variables and the collected data, to obtain ideas for problem solving and develop a plan.

Third, implement the plan. The plan created in the second step is implemented.

Fourth, review, examine, and discuss the answers.

(5) Impact assessment

Considering differences in students' problem-solving ability, we select appropriate questions from the question database to evaluate the effect of cognitive diagnosis. If the effect is not satisfactory, we need to design another typical problem to conduct cognitive diagnosis and intervention for students.

7.3.2 *Role of the Cognitive Model*

Through surveying teachers, we found that primary mathematics teachers' diagnosis of the causes of students' problem-solving errors depends on their personal experience and intuition. These diagnoses lack objectivity and scientific support. The cognitive model can assist with cognitive diagnosis.

(1) Provide a reliable basis for cognitive diagnosis

Given that medical diagnosis requires standard data, the diagnosis in educational contexts also requires such data. The problem-solving cognitive model CMMPS can analyze the cognitive process of problem solving and provide a reference for cognitive diagnosis.

(2) Clearly identify the cause of the problem-solving errors

By analyzing the problem-solving process through CMMPS, we aimed to capture the cognitive process of correctly answering questions and used the process as a stepping stone. To obtain the cognitive process of making problem-solving errors, we identified the internal process that led to the errors, clearly identified the causes of the errors, and then provided target intervention.

7.4 Improvement of Mathematics Teaching

7.4.1 *Paying Attention to Individual Differences in Students' Abilities*

"One-on-one" cognitive diagnosis can identify the deficiencies of the ability to solve mathematical problems for each student and can use students' incorrect performance in the process of solving problems to identify problems of different difficulty levels for students with different ability levels. When students encounter difficulties in the problem-solving process, teachers ask leading questions, gradually guide the students to provide correct solutions to the problem, respond to students' various requests and promote the sustainable development of students' mathematical abilities.

7.4.2 *Early Identification of and Intervention in Mathematical Cognitive Impairment*

Through "one-on-one" cognitive diagnosis, it is proposed to identify learning difficulties in advance and reduce or eliminate them through corresponding remedial measures. By analyzing students with learning difficulties, we aimed to identify the cognitive barriers that lead to learning difficulties for different types of questions and various grades and analyze the causes of cognitive impairment. As the famous saying goes, "Rome wasn't built in a day"; the cognitive obstacles of students in higher grades may gradually accumulate from learning in the lower grades. Therefore, cognitive impairment should be prevented and interventions should be implemented in the lower grades, thus generating a positive impact on learning. For example, consider a student who has a problem with two-digit multiplication. A careful analysis of the student's calculation process could enable the teacher to know that the student has mastered the simple rules of multiplication but makes mistakes in "carrying the addition," which is to say, the student has formed a certain cognitive model in his or her brain. To solve this problem, when teachers teach first graders the calculation method for the first time, they should guide students to pay extra attention to carrying the addition, apply an instructional method that is suitable for students' cognitive characteristics and be patient when providing explanations. By doing so, teachers can effectively prevent potential learning problems that may arise in future learning with minimal effort.

7.4.3 *Specific Guidance for Students with Cognitive Impairment in Mathematics Learning*

An analysis of the cognitive impairment of students with learning difficulties revealed that similar or identical cognitive impairments exist within the same grade. To conduct an in-depth analysis of the typical characteristics of cognitive impairment, we identified the reasons for the latter and formulated effective interventions to provide targeted guidance for students with typical cognitive impairment. To change the current situation of "one set of instructions for all students," targeted guidance, on the one hand, can save students' time and increase their interest in learning mathematics, and, on the other hand, can target specific mathematical problems with in-depth explanations and thus apply methods depending on the situation. It is a good choice to conduct "one-on-one" cognitive diagnosis in schools with good teaching conditions.

7.4.4 Making Good Use of Students' Zone of Proximal Development to Promote Cognitive Development

Teachers can learn from the diagnosis of students' cognitive level, use education goals and objectives as a reference, formulate a series of intervention strategies, present them to students in the appropriate sequences, and consciously participate and intervene in the learning process. In this process, students acquire knowledge, skills and problem-solving strategies, internalize them into their original cognitive structure, form new cognitive structures, and further promote their cognitive development. Feuerstein noted that the experience of intermediary learning accompanies the process of growing up. The quality of intermediary learning directly affects the cognitive development of individuals.

7.5 Summary

This chapter discusses the application of problem-solving cognitive process analysis in teaching, i.e., "one-on-one" cognitive diagnosis and intervention. This chapter first defines the concept and theoretical basis of cognitive diagnosis. It then analyzes the cognitive diagnosis that was conducted on teaching primary school math problems and proposes "one-on-one cognitive diagnosis." This chapter explains the purpose and characteristics of this diagnostic method, emphasizes the pertinence and timeliness of intervention, proposes a method and process of diagnosis, and finally discusses the implications of cognitive diagnosis for mathematics education.

Studies abroad have been actively examining cognitive diagnosis. Many theoretical and applied studies have been carried out and have achieved satisfactory results. In China, research on cognitive diagnosis is in contrast relatively limited, and the majority of such studies are introductory and theoretical, which can hardly be used to guide practical teaching. Even fewer empirical studies examine the cognitive diagnosis of mathematical problems, especially in primary schools, which should be further studied.

References

Anand, P. G., & Ross, S. M. (1987). Using computer-assisted instruction to personalize arithmetic materials for elementary school children. *Journal of Educational Psychology, 2*, 72–78.

Campbell, C., & Carlson, J. S. (1995). The dynamic assessed of mental abilities. *European Distribution on Dynamic Assessment, 3*, 1–31.

Chen, Y. H., Na, G. L., & Zhong, N. N. (2004). Mathematical problems on the use of mental representation of the new theory. *Psychological Science*, 1.

De Smedt, B., Ghesquiere, P., Swillen, A., & Verschaffel, L. (2009). Mathematical learning disabilities in children with 22q11 2 deletion syndrome: A review. *Developmental Disabilities Research Reviews, 15*, 4–10.

DiBello, L. V., & Stout, W. (2007). "Guest editors" introduction and overview: IRT-based cognitive diagnostic models and related methods. *Journal of Educational Measurement, 44*, 285–291.
Geert, P. V. (1998). A dynamic systems model of basic developmental mechanisms: Piaget, Vygotsky, and beyond. *Psychological Review, 105*, 634–677.
He, H. D., & Fu, X. L. (1995). A process of problem characterization. *Acta Psychologica Sinica.*
Kane, M. T. (1992). An argument-based approach to validity. *Psychological Bulletin, 112*, 527–535.
Kaufmann, L., Kremser, C., Wood, G., Schocke, M., Starke, M., & Vogel, S. E. (2009). Developmental Dyscalculia: Compensatory mechanisms in left intraparietal regions in response to nonsymbolic magnitudes. *Behavioral and Brain Functions, 5*, 35–40.
Kotovsky, K., Hayes, J. R., & Simon, H. A. (1985). Why are some problems hard? Evidence from Tower of Hanoi. *Cognitive Psychology, 17*, 248–294.
Lane, S. (2004). NCME presidential address: Validity of high-stakes assessment: Are students engaged in complex thinking? *Educational Measurement: Issues and Practice, 23*, 6–14.
Leighton, J. P. (2004). Avoiding misconceptions, misuse, and missed opportunities: The collection of verbal reports in educational achievement testing. *Educational Measurement: Issues and Practice, 23*, 6–15.
Leighton, J. P., & Gierl, M. (2007). Verbal Reports as Data for Cognitive Diagnostic Assessment. In J. P. Leighton & M. Gierl (Eds.), Cognitive Diagnostic Assessment for Education: Theory and Applications (pp. 146–172). Cambridge University Press.
Mayer, R. E. (1987). *Educational psychology: A cognitive approach.* Little, Brown.
Messick, S. (1989). *Educational measurement* (3rd ed.). American Council Education/Macmillan.
Ministry of Education of the People's Republic of China. (2022). *The compulsory education mathematics curriculum standards* (2022 edition). Beijing Normal University Press.
Montague, M. (1992). The effects of cognitive and metacognitive strategy instruction on mathematical problem solving of middle school students with learning disabilities. *Journal of Learning Disabilities, 25*, 230–248.
Shi, T. (1985). Cognition model of algebraic application problem. *Acta Psychologica Sinica.*
U.S. Department of Education. (1965). *Elementary and secondary education act.* http://www.ed.gov/esea.
U.S. Department of Education. (2001). *No child left behind.* http://www2.ed.gov/nclb/landing.jhtml.
Yang, X., & Embretson, S. E. (2007). Construct validity and cognitive diagnostic assessment (pp. 119–145). Georgia Institute of Technology.

Chapter 8
Implications of "One-on-One" Cognitive Diagnosis

8.1 Experimental Design

The "one-on-one" cognitive diagnosis and intervention experiments include the following sections.

(一) Preknowledge test

This test is used to judge students' mastery of the prerequisite knowledge required for learning new knowledge points. Preknowledge testing also tests some of the new knowledge points. An additional test is given to students who answer the questions correctly to determine whether the student has already mastered the new knowledge to ensure that learning occurred in the experimental process.

(二) Typical problem design

Typical problems are designed to test representative knowledge points according to the daily lives of students. These problems include preclass inquiry and after-class inquiry.

(三) Preclass inquiry

Preclass inquiry questions are assigned to students to be completed independently one day before the new class is taught. The oral reporting method is employed to record students' problem-solving process. Data received from oral reports are translated and used for analyzing the cognitive process of students' problem solving and mistakes made in the process and for formulating an appropriate intervention plan.

(四) After-class study

After class, all planned problems are taught, and students' mastery of new knowledge is examined through inquiry. The oral reporting method is used to record students' process of analyzing and solving problems.

X. Wei, *Artificial Intelligence in Mathematics Education*, Smart Computing and Intelligence, https://doi.org/10.1007/978-981-95-1405-2_8

(五) "One-on-one" cognitive diagnosis intervention

Some students who incorrectly solve problems are selected for a "one-on-one" diagnosis and intervention. For students whose cognitive diagnosis intervention is unsuccessful, a second cognitive diagnosis intervention is implemented.

The "one-on-one" cognitive diagnosis and intervention process is shown in Fig. 8.1.

8.2 Empirical Study of "One-on-One" Cognitive Diagnosis of Declarative Knowledge

8.2.1 Purpose

The concept of "mode" is a knowledge point in the sixth unit, "statistics", of the math textbook for fifth graders (Course Textbook Institute, 2006). It is a typical example of declarative knowledge. The purpose of the experiment is to carry out "one-on-one" cognitive diagnosis and intervention for students who have difficulties with declarative knowledge problem solving and then analyze and compare the changes in students' problem-solving abilities before and after the intervention.

8.2.2 Methods

(一) Subjects

A total of 28 students in the fifth grade of Yongliang Primary School in Gaoyang County, Hebei Province, were selected as subjects. Of the students, 10 were boys and 18 were girls, with an average age of 134 months and an age range of 129–143 months.

(二) Materials

"Mode" application questions in fifth-grade math are examples of typical declarative knowledge and represent a new knowledge point for students who have mastered the required prior knowledge. The application of "mode" is presented in the form of a "birthday" situation, in which students are closely connected with their daily life. They are given the role of "the classroom teacher". Through solving the actual problem, students are expected to summarize the method of determining the "mode" (students do not know that they are being asked to provide the "mode"). The researchers then diagnose any problems that students exhibit in determining the mode.

The test materials are designed for the purpose of this study and are as follows:

1. Preclass knowledge test

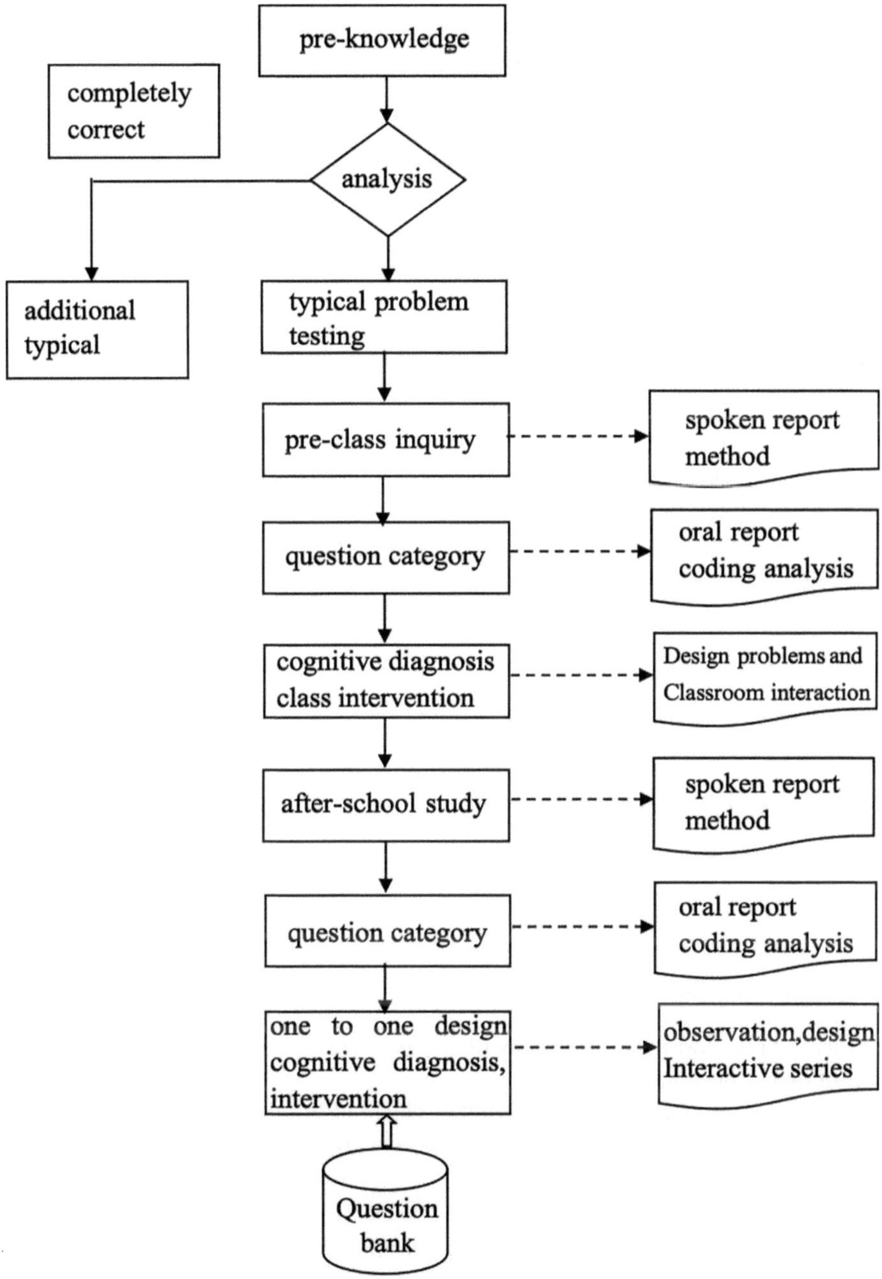

Fig. 8.1 "One-on-one" cognitive test procedure

(1) China has a profound history and culture of surnames, and each surname has a unique and rich cultural connotation. Imagine that you want to rank the surnames in our class:

① How would you collect students' surnames?
② Count the number of occurrences of each surname.
③ Arrange the surnames in descending character stroke order.

(2) Find the average and the median of the numbers listed below and show your calculation process.

3.6, 2.4, 2.8, 2.9, 3.2, 2.1, 2.2

(3) Please fill in the blanks according to the 25 numbers given.

1.32	1.33	1.33	1.33	1.33	1.33	1.33	1.44	
1.45	1.46	1.46	1.47	1.47	1.48	1.48	1.49	
1.50	1.51	1.51	1.52	1.52	1.53	1.54	1.55	1.55
1.32 is (minimum)								
1.55 is ()								
1.47 is (median)								
1.33 is ()								

2. Pretest additional questions

Think about the "mode" problems in everyday life and provide reasons.

3. Oral report training questions

The oral report training questions are shown in Fig. 8.2.

4. Preclass inquiry

Please brainstorm and come up with a good idea to help your head teacher with the following question:

Our class is going to have a birthday celebration next year, but we can only celebrate students who were born in a certain month. Imagine you are the head teacher:

(1) What factors would you consider?
(2) How are you going to choose the month?
(3) Which month should be chosen? Why?

5. After-class inquiry

A shoe factory plans to promote a batch of new sports shoes, and the factory intends to provide shoe samples of a certain size for free trial to our students. If you are the student representative, how would you choose the shoe size?

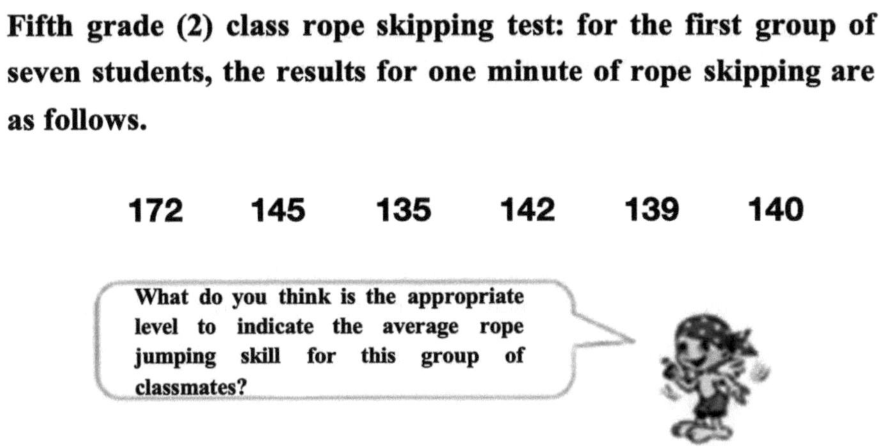

Fig. 8.2 Oral report training questions

6. The first diagnostic intervention topics

   (1) Data collection topics

The visual acuity of the left eye of Grade 5, Class 1 students is as follows:

5.0	4.9	5.3	5.2	4.7	5.2	4.8	5.1	5.3	5.2
4.8	5.0	4.5	5.1	4.9	5.1	4.7	5.0	4.8	5.1
5.0	4.8	4.9	5.1	4.5	5.1	4.6	5.1	4.7	5.1
5.0	5.1	5.1	4.9	5.0	5.1	5.2	5.1	4.6	5.0

What is the mode of this set of data?

(2) Strategy topics

The entire school needs to choose one place for a spring tour, and the available options are as follows:

A. Beijing Tiananmen Square B. Baiyangdian C. Taishan Mountain

If you were the principal, how would you choose?

7. Second diagnostic intervention topics

   (1) Data collection topics

A shooting team needs to select one of two athletes to participate in a competition. They each play 10 rounds of ammunition. The results are as follows:

A:	9.5	10	9.3	9.5	9.6	9.5	9.4	9.5	9.2	9.5
B:	10	9	10	8.3	9.8	9.5	10	9.8	8.7	9.9

① What is the average and mode of each athlete's scores?
② Who do you think should participate in the competition? Why?

(2) Strategy topics

The school plans to hold a New Year's party. Each class can perform only one show, but our class has three candidate programs:

① The solo song "Where is spring" by Xiao Ming
② The solo dance "We are the flower buds" by Xiao Hong
③ A poetry recitation of "Facing the Ocean, Spring and Blossom" by Xiao Lan

For the sake of fairness, if you were the classroom teacher, what would you do?

(三) Procedures

1. Design an experimental program

According to the purpose of the experiment, the experimental subjects, materials and instructions are finalized, the overall experimental process is determined, and the experimental results are obtained.

2. Experimental equipment

A single recording pen, a Sony video camera, and a tripod were used to record the data generated from the oral reports.

3. Pretest

After the prerequisite knowledge required to solve the "mode" problem was analyzed, suitable questions were designed and applied to test students' understanding of the prerequisite knowledge. Students who correctly answered questions in the pretest were asked to answer the extra set of questions to determine whether they had mastered the "mode" concept.

4. Verbal reports and records

The verbal reporting method was applied to collect the data. In accordance with the think aloud protocol developed by Erickson and Simon, the participants were trained to verbally convey their thinking process through the problem-solving process. Before starting the question, the researcher first briefly explained the guidelines and then used an example to demonstrate how to thoroughly reflect the thinking process during the problem-solving process. After the participants confirmed their understanding, they started to answer the questions, and a video camera was used for recording the process.

5. Data translation and coding

The data collection included two parts: a verbal report and video recording of the problem-solving process. For the verbal report data, the translator first dictated the audio files into the texts and then combined the students' problem-solving assignments to conduct coding analysis to diagnose learning difficulties. The coding work was undertaken by two professionals, and disagreements during the coding process were resolved through discussions.

6. Preparation of intervention programs

According to the collected verbal reports and problem-solving videos, we analyzed the students' thinking and obstacles in the process of solving problems and formulated specific intervention plans, such as tips to help students solve problems and obtain better performance.

7. Interview

An interview was conducted with the teachers in the course of the students' problem-solving process to understand the typical performance of the students and to refine the intervention programs.

8. Intervention

An intervention program was used to address students' difficulties in the problem-solving process. The participants were given sufficient time to work on the questions, and their responses were observed.

9. Effect evaluation

The evaluation of students' knowledge mastery was conducted through a questionable design.

(IV) Experimental duration

The dates of the experiment were December 12–15, 2011.

8.2.3 *Results Analysis*

Throughout the experiment, 28 students in 5th grade participated in the preclass inquiry question verbal report experiment. All the students participated in the after-class inquiry question verbal report experiment, 18 students participated in the first cognitive diagnosis intervention, and 13 students participated in a second cognitive diagnosis intervention. Over the course of two cognitive diagnosis interventions, each student was provided two categories of questions: data statistics and strategic solutions. A total of 118 verbal reports were obtained during the experiments.

(一) "One-on-one" cognitive diagnosis of the overall effect analysis

Table 8.1 shows the statistics of the pretest results for the "model"

In the table, 1 (1) represents the first question in the pretest, 1 (2) represents the second question in the pretest, and so on. The statistical results revealed that, for the first question, 1 (1), which examined how the students collected data, 8 students adopted the method of writing the last name on paper first and then calculating the statistics, and the strategy was clear. However, the other students used direct

Table 8.1 "Mode" prerequisite knowledge test results

	1 (1)	1(2)	1(3)	2(1)	2(2)	3	4
The correct number of people	8	23	20	25	26	9	2
Correct rate (%)	28.6	82.1	71.4	89.3	92.9	32.1	7.1

counting. 1 (2) shows the results of the surname statistics; five students missed a few surnames and made a statistical mistake. 1 (3) tested the surname arrangement; 8 students incorrectly arranged the surnames and therefore made some mistakes. The analysis revealed that the choice of statistical strategy was more difficult than the specific statistical process. For question 2, 3 students made mistakes on question 2 (1) "average" and 2 students made mistakes on question 2 (2) "the median question". The median was easier to compute than the average was because the latter involved summation and division calculations, and both calculation processes were prone to errors. Question 3 examined students' understanding of the learned concepts; the accuracy rate was lower than one-third. Nine students filled in "the number appeared most" (not knowing that "the number appeared most" is the "mode"). Question 4 is an additional test that was given to the students who answered the previous questions correctly. It examines whether students understand the concept of "mode". Only two students provide situational examples; for example, the first digit of a phone number in China is a "mode", which is 1, and the number of students who answered correctly is 7.1%. The above results indicate that the prerequisite knowledge required to learn the mode was mastered by the majority of the students, whereas the concept of the mode was new knowledge for 92.9% of the students.

A total of 28 participants participated in the preclass exploration phase. Among them, MaYY, DuanYM, and WangYF answered the questions correctly, accounting for 10.7% of all participants. WangWY and DuanJN understood the meaning of the questions but lacked a comparative strategy. Specifically, they did not know how to compare the numbers of people who had a birthday in each month and accounted for 7.2% of the participants. MaYP, ZhangN, DuanYC, DuanZX, DuanHJ and CuiW did not understand the meaning of the "classroom teacher", accounting for 21.4% of the participants. LiuZ, LiuML, DuanQ, XingQL, ChengYN, WangC, and 14 other students obtained no solution to the problem, accounting for approximately 50% of the participants. Verbal report analysis revealed that most students did not solve the problem with only one strategy. In the course of problem solving, the strategy of solving the problem changed.

An analysis of the verbal reports of the preclass experiments revealed that the main difficulty exhibited by the students was a lack of solutions to the questions (50%), followed by a failure to understand the questions (21.4%). Students who understood the problem and knew the problem-solving strategy but still made mistakes constitute the third category (17.9%). Students who do not understand the problem certainly lack problem-solving strategies. During the lecture or teaching process, to help students understand the instructions, teachers should draw on daily life situations, emphasizing the "birthday" situation and the role of the "classroom teacher". To address the statistical errors, teachers should invite the students who make mistakes to count "birthdays" in class. After clarifying the instructions, the research proceeded to the inquiry stage.

A total of 27 students participated in the after-class inquiry phase, among whom 17 students (LiuZ, DuanYD, DuanZX, Du-anJN, ZhangN, WangWY, DuanXY, WangSY, XingQL, DuanQ, Du-anXT, LiuJ, MaJW, ChengY, DuanYM, MaYY) correctly answered the question, accounting for 63% of the sample; WangC, LiuML, MaHR, DuanXN made statistical errors, accounting for 14.8% of the sample; DuanYC, ChengYN did not present a clear problem-solving strategy, accounting for 7.4% of the sample; and MaYP, DuanHJ, WangYF, and LiuYR did not understand the meaning of the question, accounting for 14.8% of the sample. Table 8.2 shows the results of the survey and after-school research.

As shown in Table 8.2, after the classroom intervention, the students experienced different degrees of improvement in problem-solving comprehension, problem-solving strategies and statistics, which shows that learning had taken place and that the intervention was effective.

For students who did not correctly answer the questions, "one-on-one" cognitive diagnosis and intervention were applied.

(二) Changes in the cognitive processes before and after the diagnosis of students' problem-solving difficulties

To compare the changes in students' problem-solving performance before and after the intervention, WangC is used as an example. We analyzed the changes in this student's cognitive process through the use of verbal report data.

1. Preclass inquiry stage

The date of the preclass inquiry was December 12, 2011.

The CMMPS-based "mode" preclass inquiry problem-solving cognitive process analysis is shown in Table 8.3.

Each column represents the content of a module at different times in the problem-solving process. The leftmost column of numbers indicates the row number, and each row represents a cognitive logic step, which is not the same as the actual problem-solving step. The last row indicates the end of the process, that is, the completion of the problem-solving process.

The results of the "mode" verbal inquiry analysis of WangC's pretest are shown in Table 8.4. The left side of the table shows the verbal report data recorded during

Table 8.2 "Mode" knowledge question, preclass and after-school survey results

	Correctly answered the question	Did not understand the problem	Problem solving strategy was not clear	Statistical error
Preclass study (%)	10.7	21.4	50	17.9
After-school exploration (%)	63	14.8	7.4	14.8

Table 8.3 "Mode" preclass inquiry question problem-solving cognitive process analysis

	Visual	Production	Retrieval	Goal	Imaginal	Manual
1	Text encoding					
2			Relevant semantic knowledge in DM			
3				Which month should be chosen?	Which month should be chosen?	
4		Choose a month to celebrate students' birthdays, assume the class teacher role → select the month with the most birthdays (in line with common sense)				
5				Select the month with the most "birthdays" (target conversion)		
6						Select the month that has the most number of students with birthdays, and then choose that month
7					Select the month that has the highest number of students with birthdays	
8				Statistics on the number of student birthdays in a month		

(continued)

Table 8.3 (continued)

	Visual	Production	Retrieval	Goal	Imaginal	Manual
9			Statistics, counting			
10					Compare the number of birthdays each month	
11			The comparison of size of the number			
12				Choose the month with the largest number of "birthdays"		
13						The "month" with the largest number of birthdays
14						Finish

the experiment. The right side shows the cognitive process for oral report data and CMMPS analysis. The diagnosis is based on a comparison between the cognitive process analysis and the cognitive process (as shown in Table 8.3) of the "mode" problem solving based on CMMPS.

2. Inquiry stage after class

The date of the after-school inquiry was December 13, 2011.

The CMMPS-based "mode" after-school inquiry problem-solving cognitive process analysis is shown in Table 8.5.

Each column shows the contents of a module at different times in the problem-solving process. The leftmost column of numbers indicates the row number. Each row represents a cognitive logic step, which is not the same as the actual problem-solving step. The last row indicates the end of the cognitive process, that is, the completion of the problem-solving process.

WangC's after-class inquiry oral report is shown in Table 8.6. The left side of the table shows the oral report data recorded during the experiment. The right side is the cognitive process for the oral report data and CMMPS analysis. The diagnosis results compares the cognitive process analysis on the right side with the "mode" after-class inquiry based on the CMMPS problem-solving cognitive process (as shown in Table 8.5).

3. The first cognitive diagnosis and intervention

Table 8.4 Oral report analysis of WangC's preclass results for the "mode" question

Oral English report	Cognitive process analysis
[Reading] Our class will hold a birthday celebration next year, but only students whose birthdays are in a certain month will be celebrated. If you are the classroom teacher, which month would you choose? Why? [Analysis] I will celebrate a birthday for MaJW [Q: Why?] Because MaJW is my best friend [Q: If you choose MaJW, which month would you choose?] March [So, you think that March is a suitable selection?] Suitable [Q: This birthday celebration takes into account all your classmates. Why March? Is it suitable?] Because March covers the most birthdays [Q: How do you know it is March?] Other people told me [Q: Who are "other people"?] WangYF [Q: What choice do you think is more appropriate? The month of your best friend's birthday or the month that accounts for the birthdays of the most students?] The most [Q: Why?] Because I do not know which month is my friend's (which month their birthday is in) [Q: Which month would you choose as your decision?] August [Q: Why?] Because there are many birthdays in August [Q: How are you going to choose the month?] [Q: A lot of birthdays are in March, and a lot of birthdays are in August. Which month would you choose then?] March [Q: If you want to know which month covers the most birthdays, how would you determine the answer?] Check on the computer [Q: How would you check it?] Baidu, enter what you want to check [Q: Do you think Baidu can search for which month has the most birthdays?] Can't [Q: What do you want to do?] Ask classmates [Q: How many?] Ask two [Q: Why?] Because if one does not know, you can ask another [Q: What if neither of these two people knows?] Ask someone else [Q: Who?] MaJW [Q: In addition to celebrating the month of your best friend's birthday, for which other month do you want to celebrate birthdays?] March [Q: Why?] Because March (the 3rd month, the number 3) is singular [Q: There are a lot of other singular numbers, why March?] Singular represents boys [Q: Which month do you want to choose then?] March [Q: Why?] Because March is singular [Q: There are many single months?] There are more people (with birthdays) in March [Q: Do you think this problem is difficult?] A little bit	Input text information by reading the title, perform visual coding to form a propositional text frame and question patterns (Give the answer directly) The basis of the choice, which comes from the content of the Production module, is "select a friend's birthday month" (Guide student to reflect on whether their answer is appropriate) (Emphasize the problem situation) Switched strategy, Production module content changed to "select the month with the largest number of birthdays" (Guide student to identify problem-solving strategies) Determine the problem-solving strategy, that is, the contents of the Production module become "select the month with the largest number of birthdays" Highlight the contradiction (Inconsistent with the previously identified problem-solving strategy) (Say it to oneself) (With computer) (Previously used Baidu, conducted a search) Production module contents become "a good friend's birthday" Solving the problem-solving strategy again, the contents of the Production module become "the month with the largest number of birthdays" Problem-solving strategy, the content of the Production module becomes "select a friend's birthday and the month with the largest number of birthdays"

(continued)

Table 8.4 (continued)

Oral English report	Cognitive process analysis
[Diagnosis] WangC students shows strong divergent thinking ability. He responded quickly and came up with a variety of strategies and methods, but his problem-solving strategy was not clear, and he did not provide a correct answer to the question	

Note () in the content is for description and explanation

The first cognitive diagnostic intervention was conducted on December 14, 2011.

The first cognitive diagnosis and intervention included two parts: data statistics and strategy selection categories. The former was used to diagnose and intervene in problems with statistics in the problem-solving process; the latter was used to diagnose and intervene in difficulties in the process of problem solving.

Table 8.7 shows the diagnostic interventions for the data statistics. The left side of the form Shows the oral report data recorded during the process, and the cognitive process for the oral report data obtained via CMMPS analysis is on the right.

WangC's strategy selection class questions for the cognitive diagnosis intervention oral report are shown in Table 8.8. The left side shows the oral report data recorded during the experiment. The right shows the cognitive process of the oral report data obtained via CMMPS analysis.

4. The second cognitive diagnosis intervention

The second cognitive diagnosis intervention was conducted on December 15, 2011.

The second cognitive diagnosis intervention also included two parts: data statistics and strategy selection. The former was used to diagnose problems in the statistical process and provide an intervention; the latter was used to diagnose strategy selection problems and provide an intervention.

Table 8.9 shows the diagnostic intervention experienced by WangC. The left side of the table shows the oral report data recorded during the experiment. The right side is the cognitive process analysis obtained from the CMMPS analysis of the oral report data.

The oral report data of WangC's strategy selection questions for the cognitive diagnosis intervention are shown in Table 8.10. The left side of the table shows the oral report data recorded during the experiment. The right side is the cognitive process analysis obtained from the CMMPS analysis of the oral report data.

5. Comparative analysis of different stages of the problem-solving process

For the purpose of comparison, Table 8.11 presents the changes in the solving process in the four stages of preschool inquiry, after-school inquiry, first cognitive diagnosis intervention and second cognitive diagnosis intervention.

As seen from the comparative analysis of Table 8.11, the significant changes in WangC's problem-solving process before and after the interventions helped him answer the questions correctly and develop an effective way of solving problems.

Table 8.5 "Mode" inquiry question problem-solving cognitive process analysis

	Visual	Production	Retrieval	Goal	Imaginal	Manual
1	Text encoding					
2			Relevant semantic knowledge in DM			
3				How would you choose the shoe size?	How would you choose the shoe size?	
4		Choose a shoe size, assume the monitor role → choose the shoe size that the most students wear (in line with common sense)				
5				Choose the shoe size that the most students wear (target conversion)		
6						Which shoe size do the most students wear, and then choose that shoe size
7					Which shoe size do most students wear	
8				Statistics for each student's shoe size		
9			Statistics, counting			

(continued)

Table 8.5 (continued)

	Visual	Production	Retrieval	Goal	Imaginal	Manual
10					Compare the number of students in the class with each shoe size	
11			Comparison of the sizes of the numbers			
12				Choose the shoe size of the largest number of students		
13						The shoe size of the largest number of students
14						Finish

Table 8.6 Oral English report after analysis of the after-class "mode" inquiry by WangC

Oral English report	Cognitive process analysis
[Read] A shoe factory is promoting a number of new sports shoes. The shoe factory intends to provide free trials to our classmates. If you are the classroom teacher, how would you choose the shoe size?	Input text information by reading the title, and form a propositional text frame and question patterns through visual coding

(continued)

Table 8.6 (continued)

Oral English report	Cognitive process analysis
[Analysis] I choose size 39 because my shoe size is 39. Choice No. 39 is suitable [Q: Look at the question. Is this choice appropriate?] Yes [Q: If you were the classroom teacher, would this choice be appropriate?] Inappropriate, I should choose the size more people can wear [Q: So how do you know which shoe size is the most popular?] By raising hands; it is easier to raise your hand (Teacher gives the student a list of the shoe sizes of the entire class) [Q: Now you have data. What are you going to do?] (counting the number of students with each shoe size) Choose 37 because 37 covers the most students. There are eight; 38 and 39 are both 7, according to my counting, and there are 2 people with size 40 [Q: Do you think your counting is right or wrong?] I think I am right because I think I counted carefully [Q: Did you count this quantity correctly?] Yes, oh no, missed one. I thought there is no size 36. I started from size 37. There are 4 in size 36 [Q: Which one is your choice?] I choose size 37 because 37 covers the most people. The factory could provide 8 pairs of shoes [Q: Do you think this question is difficult?] A little [Q: Compared with the birthday question?] This one [Q: Why is that? It didn't take long for you to figure it out] I thought for a moment [Q: Do you think math is difficult?] Not difficult, as long as you listen attentively [Q: Are you good at math?] Some things good, some things bad	Determine the problem-solving strategy, Production module content is "choose one's own shoe size" (Guide student to understand the question) (Question meaning is not accurate) (Remind student of the "monitor" role) Goal module content for the "monitor" role, Production module contents become "shoe size that is worn by the most students" Manual module content: Select a method to collect data "hands up" Implementation strategy, Manual module content: Number Operation Manual module content: Select the shoe size of the largest number of students (The student is reminded of the main question tested in the problem-solving process) (Confidence in problem-solving process) Rethinking: Found errors after checking Determine the answer, Manual module content: choose the shoe size of the most students Retrieval module: extract long-term memory of the relevant knowledge (May indicate past learning inattention) (Results are unstable, can be found from the process of solving problems)
[Diagnosis] In the process of solving the problem, WangC chose a strategy from the two strategies. The student began to count the number of shoes, made an error, checked and reviewed the process, and then correctly answered the question. These results show that this student has not formed the habit of reviewing and checking their work	

Note () in the content of the paper indicates a description and an explanation

Table 8.7 First cognitive diagnosis intervention for WangC's "mode" inquiry oral report analysis (data statistics)

Oral English report	Cognitive process analysis
[Title] All the students of Class 1 of Grade 5 have had the vision of their left eye checked. The results are as follows:	Input text information by reading the title, and perform visual coding to form a propositional text frame and question patterns
[Analysis] (Starts writing on the paper: "A: The mode is 5.2") [Q: What do you understand the question to be?] What is the mode of this set of data? (Immediately crosses out the original answer. After a short period of counting, writes out on paper: Mode is 5.1) [Q: How many "5.1" are there?] 12 [Q: What about the others? How do you know that 5.1 is the most frequently occurring value?] (Writes out on paper: 7 for 5.0, 4 for 4.9, 2 in 5.3, 4 in 5.2, 3 in 4.7, 4 of 4.8, 12 of 5.1, 2 of 4.5, 2 of 4.6. $12 > 7 > 4 = 4 = 4 > 3 > 2 = 2 = 2$)	(Gives the answer directly. The title is not complete.) Incomplete characterization of visual module information (Hint: look at the problem) (After looking at the problem) Understands the question correctly. Provides the correct answer (Guide the student to explain the process of solving the problem.) Production module content: First, the number of different visual acuity values, and then compare the sizes of the numbers and select the value with the most occurrences
[Diagnosis] WangC did not read the entire question and rushed into problem solving, which led to the first incorrect answer. The teacher simply reminded the student to read the question again, and he quickly provided the correct answer	

Note () in the content indicates a description

After the experiment, WangC received a score of 94 points on the teacher-organized test. The subjects of the test questions were from "elementary mathematics grade five (People's Education Press)" and concerned the "*juvenile intelligence development*" 16 issue. The analysis of the papers revealed that WangC had no problem with the calculation; the loss of points was due to carelessness. By checking and reviewing his work, he can demonstrate clear progress.

6. Teacher Interview

With respect to WangC's usual learning behavior, we interviewed his mathematics teacher, Ms. Duan Junxiang, and she reflected as follows:

(1) WangC is very smart, outgoing, confident and active. However, poor language processing and reading skills affect mathematical understanding. In the junior grades, the student did not pay much attention to his language or reading skill development. His understanding of some questions on the application is poor, which leads to errors. As long as a question is discussed in class, he has no problem understanding the question and provides a solution, but errors may sometimes result from carelessness.
(2) WangC solves questions very quickly. His solution speed might put him in the top 2 in his class. Because he lacks a habit of checking, his error rate is higher than average.

Table 8.8 First cognitive diagnosis intervention of WangC's "mode" inquiry oral report analysis (strategy selection category)

Oral English report	Cognitive process analysis
[Read] The school has decided to organize a group tour during spring break, and there is only one place we can choose from the following: A Beijing Tiananmen Square; B Baiyangdian; C Mountain Tai If you are the principal, how would you choose the place?	Input text information by reading the title, and perform visual coding to form a propositional text frame and question patterns
[Analysis] [Question]: If you are the student representative, how would you choose? I would choose Baiyangdian [Q: Why?] Because people can go fishing at Baiyangdian, and it is my favorite place [Q: Please read the last line of the question.] I will choose the place most classmates in our class want to go [Q: Which one would you choose as the final destination then?] If I am the student representative, I will listen to my classmates; for myself, I will choose Baiyangdian [Q: Which one would you choose as the final destination then?] Listen to my classmates [Q: How did you start to think of Baiyangdian?] Because I thought I was not a monitor Features: The student started to answer the question without thoroughly understanding what the question is asking. Once the question was understood, the student immediately answered correctly	(Reads the title again) Gives the problem-solving strategy, and the Production module content: "choose my favorite place to go" (Provides hint to student) Understands the question, determines the strategy, that Production module content: "choose where the most students want to go" Strategy Selection, Production module content: Different strategies for different roles Determine strategy, Production module content: "choose the place where most students want to go" (Diagnosis of the wrong reasons due to a target set error) Goal: incorrect role play
[Diagnosis] WangC answered the question too quickly and did not thoroughly understand the problem situation (role) before solving the problem. After rereading the question and correctly recognize the assigned role, the student chose the correct strategy to answer the question	

Note () in the content indicates a description

The teacher's reflections on WangC confirm what we previously reported in verbal reports.

(三) Birthday data collection and statistical strategies in the process of problem solving

Data collection is required for solving the "mode" problem. When solving real problems, the data are not readily available and need to be collected and counted. Previous data analysis has shown that many students (e.g., DuanYC, DuanZX) have problems with data statistics, which leads to problem-solving errors. Data processing

Table 8.9 The second cognitive diagnosis intervention for WangC's "mode" inquiry oral report analysis (data statistics)

Oral English report	Cognitive process analysis
[Read] A shooting team plans to select one of two athletes to attend a competition. The two athletes each hit 10 rounds of ammunition. The results are as follows: A: 9.5 10 9.3 9.5 9.4 9.5 9.2 9.5 B: 10 9 10 8.3 9.8 9.5 10 9.8 8.7 9.9 ① What are the average and mode of the results of A and B? ② Who do you think is more suitable to attend the competition? Why?	Input text information by reading the title, and perform visual coding to form a propositional text frame and question patterns
[Analysis] Calculate the average, mode (Quickly wrote on paper) A: (9.5 + 10 + 9.3 + 9.5 + 9.6 + 9.5 + 9.4 + 9.5 + 9.2 + 9.5)/10 = 95/10 = 9.5 A: The average is 9.5, and A: The mode is 9.5 B: (10 + 9 + 10 + 8 3 + 9 8 + 9 5 + 10 + 9 8 + 8 7 + 9 9)/10 = 95/10 = 9.5 A: The average is 9.5, and A: The mode is 10 A: I think A should attend because (A and B have the same average) shot higher than 9 in each attempt	(Correct understanding of the problem) (Output problem-solving process) Manual module content (Correct calculation) (Output problem-solving process) (Correct calculation) Manual module content (Thought through fully)
[Diagnosis] WangC calculated the answer accurately and solved the question quickly. The student chose the candidate logically when conducting the solving process, and the student's understanding was comprehensive	

Note () in the content indicates a description

for the mode includes collecting the number of birthdays each month and determining the month with the highest number of birthdays after computing the statistics. Figure 8.3 further reveals the image, data collection, statistical methods, and preparation of the ACT-R simulation program.

The figure shows the number of people with birthdays in May (denoted by M), that is, the number of M's. The strategy adopted is to count the number of M's in the first row, then the number of M's in the second row, and finally the number of M's in the third row. The output is 5, which means that "the number of students whose birthday month is May is 5". Figure 8.4 visualizes this data collection process, with the red circles denoting the final numbers. The visualization of this strategy helps students address the "impossible" challenge of facing the data. Of course, students can also choose a strategy according to their preference, with just one of the options mentioned above.

Table 8.10 Second cognitive diagnosis interventions for WangC oral report analysis (strategy selection category)

Oral English report	Cognitive process analysis
[Reading] The school plans to hold a New Year's Day party. Each class can perform only one program, but there are three candidate programs in our class: ① Xiaoming solo "where is spring?" ② Xiaohong solo dance "we are the flower buds" ③ Xiaolan poetry recitation "facing the sea, spring flowers" If you are the classroom teacher, to ensure fairness, what would you do?	Input text information by reading the title, and perform visual coding to form a propositional text frame and question patterns
[Analysis] I will choose the song most classmates like [Q: How do you know which one your classmates like?] Ask classmates [Q: How many students?] 28 [Q: After you ask, which one would you choose?] Which is the most, which one you choose	Correct understanding of "class teacher role", Goal module content Determine the problem-solving strategy, Production module: "selected students' favorite song" Given statistical strategy, Production module: "ask students" (Hint: ask which classmates) Production module: ask the classmates (including myself) (Hint: ask students, the next step how to do) Production module: data statistics, selection of the choice of the most students
[Diagnosis] WangC changed his previous habit of not reading the topic before attempting the problem. The student's problem-solving ideas are clear, and the problem-solving method is appropriate	

Note () in the content indicates a description

To calculate the correct statistics of the month with the largest number of birthdays, the required declarative knowledge is shown in Table 8.12.

Table 8.12 presents 10 ordinal pairs in total, where p1 indicates a pair of ordinal numbers (1, 2), p2 indicates the ordinal pair (2, 3), and so on. These ordinal pairs are used to compare the sizes of two numbers. For adults, the size of the number of comparisons has become automated. If students do not understand these pairs, they will make mistakes in comparing the numbers; that is, they will not be able to correctly select the month with the largest number of birthdays.

The number of birthdays in each month is 3 in January, 2 in February, 4 in March, 3 in April, 3 in April, 6 in May, 4 in June, 2 in July, 3 in August, 2 in September, 4 in October, 4 in November and 5 in December. Figure 8.5 shows the statistical simulation of the month with the largest number of birthdays. The result is May.

Table 8.11 Changes in WangC's cognitive process at different stages of the experiment

Stage		Problem-solving process characteristics
Before class (December 12)		Problem-solving strategy was not clear, the idea was not clear, and there was no correct answer to the question
Study after class (December 13)		The student chose from one of two strategies, started counting the number of shoes when there was an error, reviewed the strategy after the correction, and finally, provided a correct answer. The student did not display the habit of reviewing his work
The first intervention (December 14)	Data collection question	The student did not finish reading the question. He was so eager to solve the question, which led to a problem-solving error. After a simple reminder, he read the question again, quickly understood the questions, and quickly solved the questions correctly
	Strategy selection questions	The student answered the questions quickly without understanding the problem-solving strategy, which led to an error. After looking at the question again and recognizing the assigned role, he chose the correct strategy to answer the question
The second intervention (December 15)	Data collection question	Calculated accurately, answered questions quickly, considered the candidates comprehensively, demonstrated comprehensive thinking, provided correct answer to the question
	Strategy selection questions	The student changed his previous habit of starting to solve problems before understanding the question. After obtaining a correct understanding, the student's problem-solving ideas were clear, the method was correct, and the problem was solved correctly

8.2.4 *Discussion*

(一) Students' problem-solving stage is not clear

In the process of teaching, teachers focus only on the content, enabling students to accept, understand, and master the content, but do not thoroughly consider the problem-solving process and stage. Good students in the process of solving problems may be identified in terms of problem-solving steps. Students who have a general or a poor grade only accept the knowledge provided by the teacher. When new problems are solved, these students tend to recall old problems instead of analyzing new ones, often resulting in "no help at all". Helping students form a good approach to solving problems is the key to resolving their difficulties.

(二) Students' understanding of a question is incomplete, which is the most common problem in the process of solving problems.

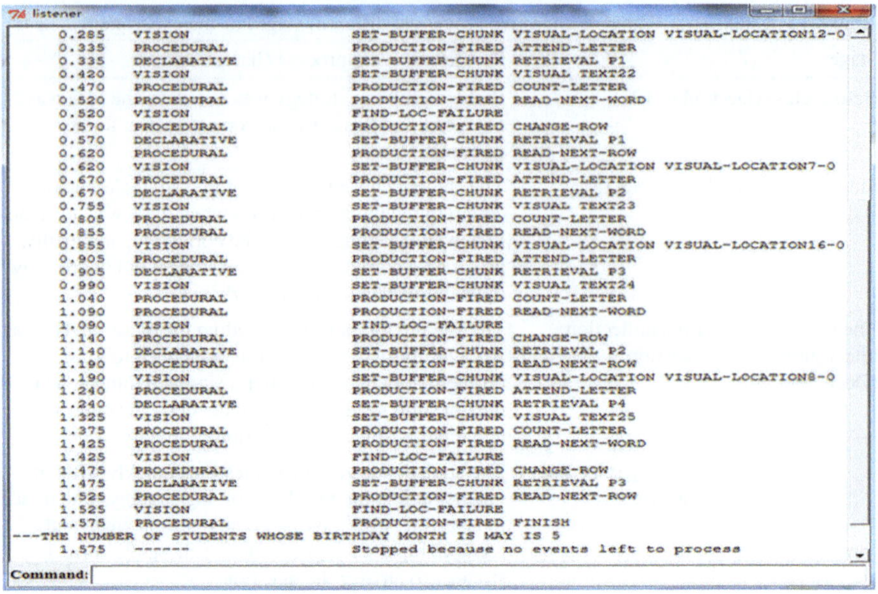

Fig. 8.3 Data collection process simulation

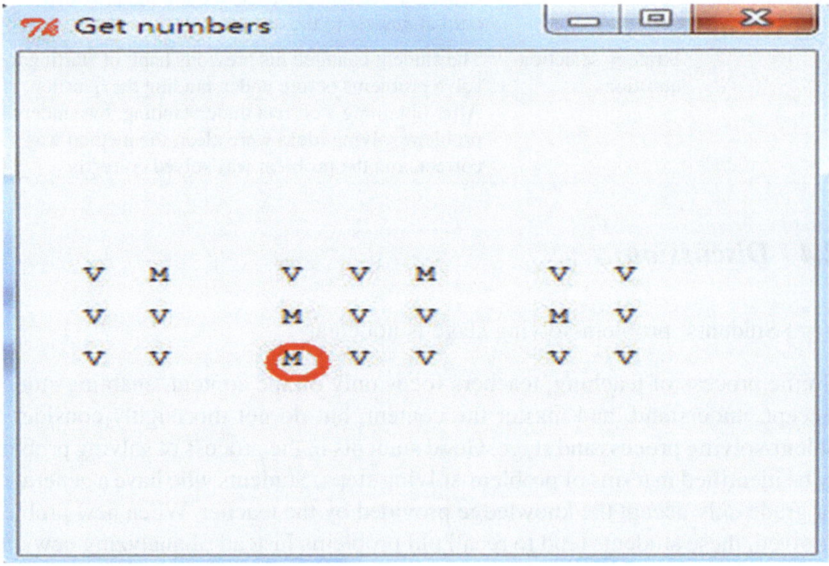

Fig. 8.4 Data collection process visualization

Table 8.12 Descriptive knowledge required for the month with the highest number of birthdays

(p1 ISA count-order first 1 second 2)
(p2 ISA count-order first 2 second 3)
(p3 ISA count-order first 3 second 4)
(p4 ISA count-order first 4 second 5)
(p5 ISA count-order first 5 second 6)
(p6 ISA count-order first 6 second 7)
(p7 ISA count-order first 8 second 9)
(p8 ISA count-order first 9 second 10)
(p9 ISA count-order first 10 second 11)
(p10 ISA count-order first 11 second 12)

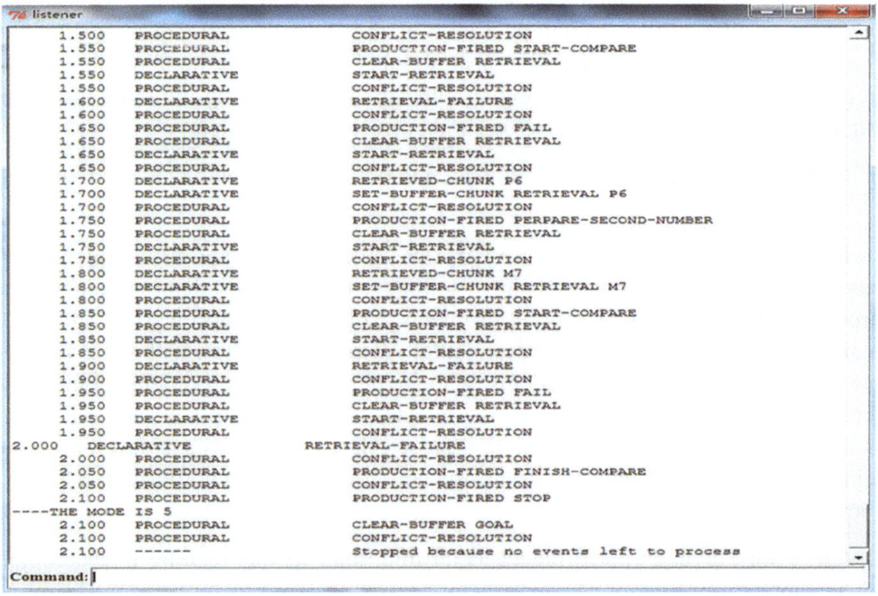

Fig. 8.5 Cognitive simulation of the month with the largest number of birthdays (part)

The above data analysis revealed that 21.4% of the students participated in the after-school research phase, and 14.8% of the students did not understand the meaning. An analysis of the results of Question 18 (Table 8.13) revealed that 15 (53.6%) students made errors because their understanding of the problem was unclear. Specifically, 5, 8, and 2 middle and poor students understood the problem. In Table 8.13, "1" means "do not understand the problem"; "2" means "understand the problem but do not know which method to use to solve the problem"; "3" means "prone to make a calculation error"; and "4" means "no habit of checking the result".

Table 8.13 Average results * difficulties encountered in solving mathematical problems

		Question 17				Total
		1	2	3	4	
Usual grades	1	5	2	6	1	14
	2	8	3	11	3	25
	3	2	4	5	0	11
Total		15	9	22	4	50

Interviews with the teacher also highlighted that the failure to understand the problem is a common problem in the process of solving problems. De Smedt's experiments involving 4th and 5th grade students found that the level of children's speech representation affected their extraction of arithmetic knowledge. In addition, the more significant the characterization was, the faster the extraction rate was (De Smedt et al., 2006). This study demonstrated that speech understanding plays a role in problem solving. After receiving simple tips for completing the main test, most students can understand the questions and correctly answer them.

(三) Students' lack of concentration is an important factor leading to an incomplete understanding of the topic.

In oral reporting experiments, researchers have reported that most students who have misinterpreted problems behave unfocusedly. Students demonstrate several problems in the process of solving problems. This phenomenon was detected later in the process of observing the experimental video. In interviews with the instructor, the teachers noted that these students, who are not attentive in problem solving, also look around and are constantly moving in class. This behavior is observed not only in math but also in Chinese and science classes. These students have developed inattentive habits, and teachers should design activities to help them develop a habit of focusing on learning.

(四) Poor language performance is an important factors leading to incomplete understanding.

Interviews with some Chinese teachers indicated that some students who did not solve the problem after the successful diagnosis also had poor Chinese scores (with a few exceptions). Poor results in Chinese can lead to an unclear understanding of the questions. It is difficult to form the correct representation of questions in the thinking process. Thus, an incorrect representation of the question is the main reason for the incorrect answers. Anand and Ross (1987) also reported that the main reason for the incorrect answer to a question was the misrepresentation of the question's structure, not the computational difficulty. Fu and He (1995) and Shi (1985) have also illustrated this point.

(五) Personal experience, family life and other background knowledge in problem solving

In the topic design part, the topic of "selecting dance partners" of the knowledge point of the PEP was analyzed. This topic was easy to understand for the students who had dancing experience. However, according to the instructor, most rural students have no dancing experience. The "selection of dance partner" problem was thus not easy to understand. In contrast, the vast majority of the students had "birthday" experience, with "birthday" situation knowledge conducive to understanding the subject. Verschaffel and De Corte (1997) also emphasized the importance of background knowledge to the application of solutions. They reported that when students solve an application problem, ignoring the real situation was not only common but also quite serious.

(六) Students solved the problem in a "nonlinear" way

There are different divisions of the stages of problem solving, but most of them describe a "linear" process that is highly rational and ideal. However, during the experiment, we found that the students did not go through all the stages of problem solving when actually solving the problem. A feedback loop is created, which is related to the development of personal habits for solving problems. In the process of solving a problem, some students look at the topic and then read the question, while others read the question first. Then, some students (including students with good academic performance, such as DuanYC and WangC) have no problem with the checking stage. On the one hand, this finding revealed the differences among the students in terms of problem-solving methods and strategies; on the other hand, it also provided a basis for the analysis of problem-solving errors.

8.3 Empirical Study on the "One-on-One" Cognitive Diagnosis of Programmatic Knowledge

8.3.1 Purpose

The "cylindrical flank area" is a part of the "cylinder and cone" knowledge taught in elementary school in the sixth grade (two volumes) textbook, Unit 2. It is an example of typical procedural knowledge in elementary school mathematics. The purpose of the experiment is to carry out "one-on-one" cognitive diagnosis and intervention for students whose procedural knowledge problem solving is difficult and analyze and compare changes before and after the intervention.

8.3.2 Methods

(一) The subjects

Fifty students from grade six of the Yong Liang Primary School in Gaoyang County, Hebei Province, were selected as subjects. Among them, 26 students were boys and 24 students were girls, with an average age of 145 months, and their ages ranged between 135 and 187 months. The proportions of males and females were basically balanced.

(二) Materials

The sixth-grade "cylindrical side area" application question is selected. This question represents typical procedural knowledge. This knowledge point is new for students, but the prior knowledge they need has already been learned. The "cylindrical flank area" application questions address the situation of "wrapping paper". The students are asked to perform hands-on operations and find the solution to the problem. For example, the "flank" can be expanded to convert the problem into a rectangular area and diagnose the problem such that the student appears in the process of solving the problem.

The test material is a test subject specially designed on the basis of the purpose of this study. The specific materials used are as follows:

1. Preknowledge test questions

   (1) The length of the basketball court is 28 m, and the width is 15 m. How many square meters is the court? How many square meters constitute the area of half the basketball court?
   (2) A rectangular vegetable garden has a long side by the wall; it is 20 m long, and 40 m are needed to fence the garden. What is the area of this vegetable garden?
   (3) A round fish pond has a diameter of 4 m. What is its area in square meters?
   (4) There is a sheep tied to a pile of wood on the grass. The length of the rope is 4 m. How many square meters of grass can sheep eat at most?

2. Oral report training problem

The oral report training exercises are shown in Fig. 8.6.

3. Inquiry questions before class

A potato chip factory produces a batch of potato chips that need to be packaged on the side of the potato chip box (as shown in Fig. 8.7). Can you help the factory calculate at least how many wrappers are needed for each chip box?

4. After-school knowledge test questions

   (1) Which of the graphics in Fig. 8.8 is a cylinder? Please mark it.
   (2) The bottom, flank, and height of the cylinder are shown in Fig. 8.9.

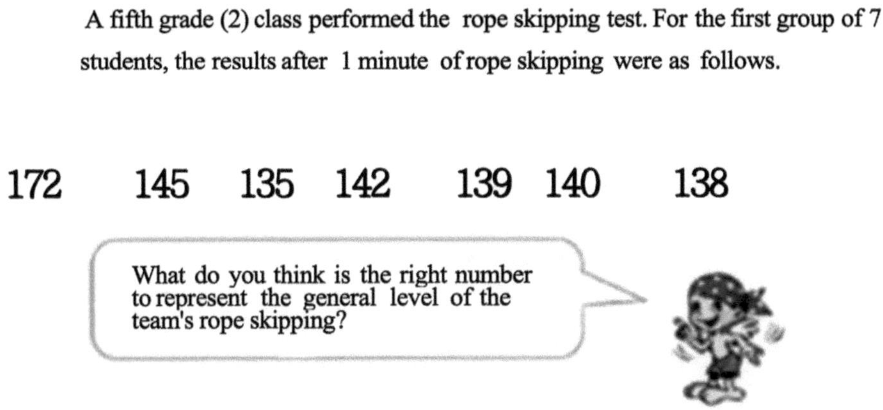

Fig. 8.6 Oral report training problem

Fig. 8.7 Chip box

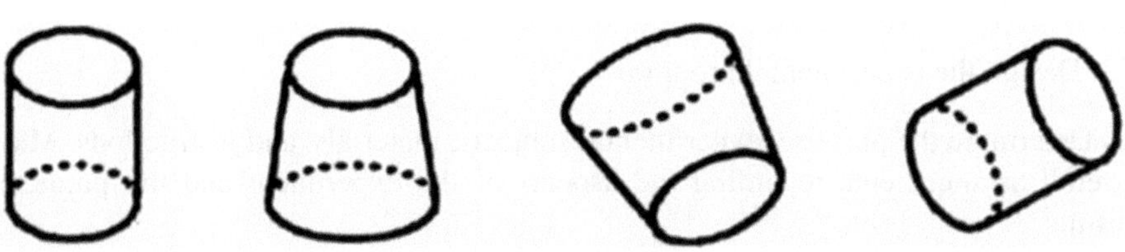

Fig. 8.8 After-school knowledge test question 1

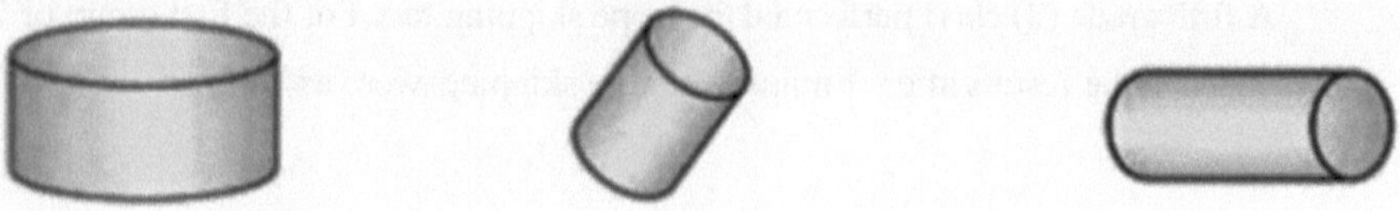

Fig. 8.9 After-school knowledge test question 2

Fig. 8.10 Roller brush

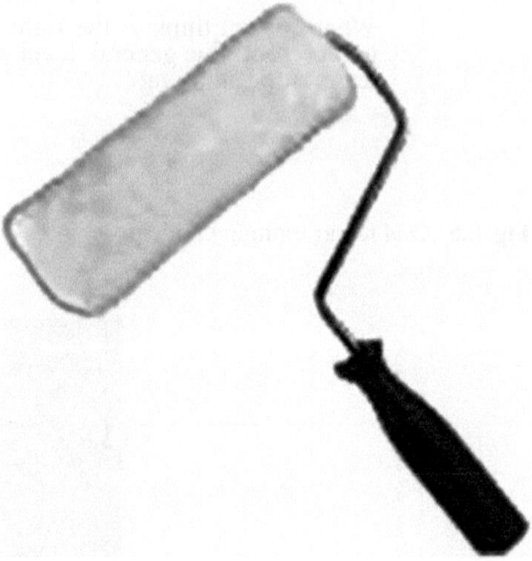

(3) A cylindrical tea box with a diameter of 5 cm and a height of 10 cm was used. Please calculate the side area of the tea box.
(4) A roller brush (shown in Fig. 8.10) is used to paint the wall; the roller brush has a radius of 6 cm and a length of 30 cm. If you dip the paint once, the roller can roll 4 turns. How many square centimeters of the wall can be brushed?

5. After-school inquiry questions

A cylinder is created after cutting a piece (as shown in Fig. 8.11). Can you calculate its side area? Note that $h = 10$ cm and $r = 4$ cm.

(三) Program

1. Design the experimental program

Determine the purpose, experimental subjects, materials, and instructions. Make overall arrangements regarding the aspects of the experiment and the predicted results.

2. Experimental equipment

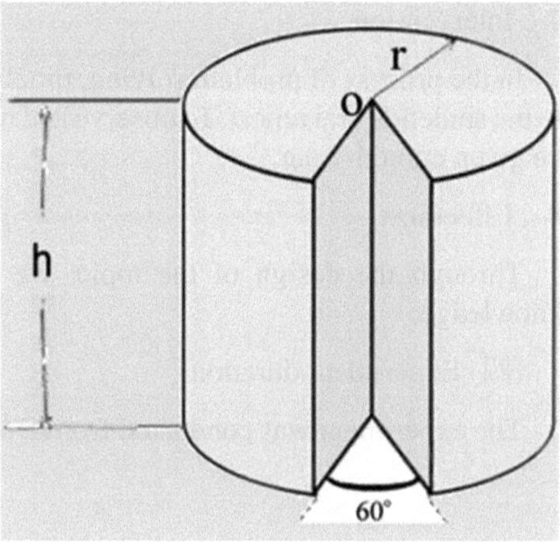

Fig. 8.11 Cylinder

A single recording pen, a Sony camera, and a tripod were used to record the oral reports of the subject during the experiment.

3. Oral reporting and recording

The oral reporting method was used to collect information. In accordance with the research methods developed by Erickson and Simon, the subjects were trained to think aloud in the process of solving the problem. Before the main question was presented, the main tester (the researcher himself) explained the requirements of the instruction. Then, the main tester used oral training questions as an example and demonstrated and explained how to think aloud in the process of answering the questions. After the subjects learned how to think aloud, they began to solve the question and were recorded via video during the process.

4. Data translation and coding

The data collected include the two parts of the oral report and problem-solving operations. The oral report is first translated into the text by the experts, and then the students' problem-solving assignments are encoded and analyzed to diagnose learning difficulties. Two experts were responsible for coding work, and agreement was reached after discussion of a small number of coding inconsistencies.

5. Development of an intervention program

The oral report data and videos collected during problem solving can be used to analyze students' ideas and obstacles in the process of problem solving and to develop specific intervention programs, such as which tips can help students solve problems and improve the learning effect.

6. Interview

The interviews with the teacher focused on the obstacles that arose in the students' problem-solving process. In addition, the students' usual performance could be fully understood, and the intervention program could be refined.

7. Intervention

In the process of problem solving, timely feedback should be provided according to the student's oral report. To observe and record students' reactions, students should be given enough time.

8. Effect test

Through the design of the topic, we evaluated the students' mastery of the knowledge.

(四) Experiment duration

The experiment was conducted from December 17, 2011, to December 22, 2011.

8.3.3 *Results Analysis*

During the experiment, the knowledge of 50 students in Grade 6 regarding the cylinder side area was tested. Forty-eight students participated in the oral report experiment of the preclass inquiry questions, and 47 students participated in the oral report experiment of the after-class inquiry questions. Forty-six students participated in the first cognitive diagnostic intervention, and 5 students participated in the second cognitive diagnostic intervention. A total of 146 students participated in the oral report experiments.

(一) The overall effect of "one-on-one" cognitive diagnosis
Preknowledge test results analysis

The statistics of the "cylindrical flank area" preknowledge test results are shown in Table 8.14.

An analysis of the students' papers revealed that the first objective was to directly examine the calculation of the rectangular area. Three students provided the incorrect answer, one student forgot to complete the second question, one person used the incorrect unit for the area, and one student made a calculation error. Question 2 was an indirect question on the area of a rectangle. Six students provided the incorrect answer because they did not understand the problem. Question 3 examined the area of the circle. Six students answered the question incorrectly, five of them did not

Table 8.14 "Cylindrical side area" preknowledge test results

	The first question	The second question	The third question	The fourth question
The number of students who solved the problem correctly	47	44	44	40
Accuracy rate	94	88	88	80

understand the problem, and one person calculated the diameter as the radius. The six students who made a mistake on the second question also made an error when they answered the third question. Question 4 was related to the actual problems of life. Ten students answered it incorrectly. Six of them did not understand the problem, three performed the calculation incorrectly, and one of them calculated the radius as the diameter. Two students incorrectly answered all four pretest questions, accounting for 4% of the sample. The test results indicate that students have mastered the preknowledge for calculating the "cylindrical flank area" of. The preknowledge test error conditions are shown in Table 8.15.

With respect to students' usual grades, even though two students usually received good grades, they made mistakes on Question 4, in which 1 did not understand the meaning of the question and 1 miscalculated the answer. Only one student who typically receives medium grades made an error in solving Question 4 due to a miscalculation. Most of the students who had poor results usually represented the majority of students who demonstrated difficulties solving the problem. According to the statistical results of previous knowledge testing, the students who were in the middle of the class in terms of grades performed slightly better than the top students did.

2. Analysis of the results of the preclass exploration stage

An analysis of the preclass research stage of 48 students' oral report data revealed that 16 students correctly answered questions, accounting for 33.3% of the sample; 13 students had clear problem-solving ideas, but their operation was incorrect, accounting for 27.1% of the sample; 4 students could not answer the question at first, but after a physical operation, they sought to calculate the area of a rectangle, accounting for 8.3% of the sample; 14 students did not know the meaning of the question, accounting for 29.2% of the sample; and 1 student did not have strategies after reading the question, accounting for 2.1% of the sample. The results are shown in Table 8.16.

The classification of errors shows that unclear understanding was the main reason leading to problem-solving errors, followed by problem-solving operations and the

Table 8.15 Preknowledge test error condition analysis

	The first question	The second question	The third question	The fourth question
Do not understand the problem		6 students (100%)	5 students (83.3%)	6 students (60%)
Mathematical formula mastery is not strong	1 student (33.3%)		1 student (16.7%)	1 student (10%)
Calculation error	1 student (33.3%)			3 students (30%)
Carelessness	1 student (33.3%)			
Total	3 students	6 students	6 students	10 students

Table 8.16 "Mode of measurement area" preclass research results

The results of preclass research	Number of students	Proportion of sample
Correct answer	16	33.3
Clear thinking, mistakes in operation	13	27.1
Cannot answer the question at first, able to do after the physical operation	4	8.3
Misunderstand the meaning of the question	14	29.2
No problem-solving strategy	1	2.1
Total	48	100

lack of a problem-solving strategy, which accounted for only a small proportion of the errors. Moreover, the physical operation helped the students solve the problem. The above statistical results can provide targeted help for the design of classroom teaching.

3. Analysis of the results of after-school research

The after-school inquiry questions were more difficult than the preclass inquiry questions were, as they examine students' understanding of the side area of a cylinder. A total of 47 students participated in the after-hour inquiry oral test report, and the statistical results are shown in Table 8.17. Three students answered the question correctly, accounting for 6.4% of the sample. Three of them adopted the remaining length of the circumference × height, the area of the side of the cylinder—the area of the side of the cylinder × 1/6 + two rectangular areas, and the three methods of bisecting the cylinder. Three students demonstrated a correct understanding of the concepts but incorrect calculations, accounting for 6.4% of the sample; 10 students demonstrated correct problem-solving ideas but were not able to calculate the cutoff area, accounting for 21.3% of the sample; 30 did not understand the meaning of the side area, accounting for 63.8% of the sample; and 1 student did not have a strategy after reading the question, accounting for 2.1% of the sample.

From the statistical results, the main reason leading to mistakes in solving problems was "do not understand which part of the side area is included"; that is, the concept of the side area was not thoroughly understood. The number of excellent, middle, and poor students each accounted for approximately one-third of the sample. For effective interventions, these students' oral reports were analyzed in depth, and the types of errors were classified, as shown in Table 8.18.

As shown in the data analysis in Table 8.18, the main reason for mistakes was "missing two rectangles", accounting for 73.3% of the sample, and the middle students accounted for nearly half. The reason why students "miss two rectangles" was that they did not understand the side area of the irregular graph. The use of a physical display during the intervention phase and the students' hands-on methods allowed the students to intuitively feel how to transform the irregular graph into a regular graph. Then, the learned side area knowledge can be used to solve the problem.

Table 8.17 "Cylindrical side area" statistics of the study results

	Correct answer	Clear thinking, mistakes in operation	Cannot answer the question at first, able to do after the physical operation	Misunderstand the meaning of the question	No problem-solving idea
Excellent	2	2	4	9	
Middle	1	1	2	13	
Poor			4	8	1
Total	3	3	10	30	1
Proportion	6.4	6.4		63.8	2.1

Table 8.18 "Do not understand side" wrong type classification

	Missing two rectangles	Miss the cutoff part	Calculate the entire side	Calculate the cutoff part
Excellent	7	2		
Middle	10	2	1	
Poor	15		2	1
Total	22	4	3	1
Proportion	73.3	13.3	10	3.4

4. Analysis of the first intervention results

An analysis of the after-class inquiry oral report information revealed that the main reason for mistakes was "missing two rectangles". In combination with examples of preclass research, such as "paper wrapping", which provides students with irregular patterns to explore, teachers provide a few tips and recall the students' problem-solving process. A total of 46 students' intervention results are shown in Table 8.19.

The intervention results revealed that 73.9% of the students correctly answered the question, with excellent and middle students accounting for the vast majority.

Table 8.19 "Cylindrical side area" results of the first intervention

	Correct answer	The idea is correct, but there is a calculation error	Do not understand which part of the side is covered	No problem-solving strategy
Excellent	17		2	
Middle	12	1	1	
Poor	5	2	5	2
Total	34	3	8	2
proportion	73.9	6.5	17.4	2.2

Eight students, including two top students, did not understand the concept of the side area. This finding shows that the two top students applied their knowledge to solve simple problems better but did not perform well when faced with complicated issues or when they were required to transfer their knowledge. Students who did not have a solution to the problem in after-class study had no solution to the problem after the intervention, and the effect was not significant.

For the 34 people who answered the question correctly, different approaches were used. The results are shown in Table 8.20.

Strategy A: cylinder side area × 5/6 + 2 rectangular area; strategy B: expansion is rectangular, long × wide; strategy C: the circumference of the underside of the residual circle is × high + 2 rectangular areas; strategy A + strategy B represents two approaches. The statistical results revealed that the majority of students adopted strategy A, accounting for 70.6% of the sample; among the students who used strategy B and strategy C, the top students had advantages, which meant that their mathematical thinking was more flexible when solving the problem.

5. Analysis of the results of the second intervention

The five students who did not understand the concept of the side after the first intervention were selected on the basis of the oral report data to diagnose the problem-solving process and perform the second intervention. Of the 5 students, 2 were excellent, 1 was in the middle, and 2 were poor. All the answers were correct. Table 8.21 lists the strategies used to generate the solution.

Strategy A: Cylindrical side area × 5/6 + 2 rectangular areas; strategy B: Circumference of remaining round bottom × height + 2 rectangular areas. The results revealed that 4 of the 5 students used the general method, strategy A; only

Table 8.20 Different strategies to obtain the correct answer to the "cylindrical side area" in the first intervention

	Strategy A	Strategy B	Strategy B + Strategy C	Strategy C
Excellent	9	3	3	2
Middle	10	1	1	
Poor	5			
Total	24	4	4	2
Proportion	70.6	11.8	11.8	5.8

Table 8.21 Different strategies to obtain the correct answer to the "cylindrical side area" in the second intervention

	Strategy A	Strategy B
Excellent	1	1
Middle	1	
Poor	2	
Total	4	1
Proportion	80	20

1 student used the problem-solving strategy of converting irregular graphics into regular graphics.

The above analysis revealed that the side area of the regular graph is easy for students to understand, which makes it difficult for students to understand the side area of the irregular graph. The key to an effective intervention is how to help students understand this concept.

With respect to the side area of the regular graphics, practice showed that the use of a physical display method and the transformation of irregular graphics into regular graphics was a good strategy after the conversion was completed, and the calculation was not difficult for the students.

(二) Changes in the cognitive process before and after the intervention are used to analyze changes in the student RanA as an example in various stages of problem solving and the cognitive process.

1. Knowledge of test question analysis

RanA been revised and correctly answered all four questions on the knowledge test, indicating that this student mastered the concepts of the area of the rectangle and the area calculation method.

2. Preclass exploration stage

The study was conducted on December 17, 2011.

The CMMPS-based "cylindrical side area" was used to obtain the cognitive process, as shown in Table 8.22.

In the process of solving the problem, two strategies are used to find the length and width of the rectangle: (1) the flank is directly expanded into a rectangle, and the length and width of the ruler are measured; (2) according to the formula of the circumference, the length of the rectangle is the circumference of the bottom circle, the width is the height of the cylinder, and the area is the circumference of the bottom circle, x, is high, that is, 2mrxh. Each column in Table 8.22 represent the content of a module at different times in the problem-solving process, and the leftmost column number represents the line number. Each line represents a cognitive logic step, which is not completely consistent with the actual problem-solving step, and the last line represents the end of the cognitive process, that is, the end of the problem-solving process. RanA's preclass oral report information on the "cylindrical flank area" is shown in Table 8.23. The left side is the oral report data recorded during the experiment, and the right side is the cognitive process, which is based on the CMMPS analysis of the oral report data. The diagnosis result is the analysis of the cognitive process based on the CMMPS analysis (as shown in Table 8.22), and a comparison is made with the oral report data.

3. After-school exploration stage

The after-school study was conducted on December 20, 2011.

The process of solving the "cylindrical side area" after-school inquiry problem based on CMMPS is shown in Table 8.24.

Table 8.22 CMMPS-based analysis of the "cylindrical side area"

	Visual	Production	Retrieval	Goal	Imagine	Manual
1	Visual coding					
2			Relevant semantic knowledge in long-term declarative memory			
3				At least how much wrapping paper is needed?	At least how much wrapping paper is needed?	
4	Coding at least how much wrapping paper is needed?					
5		At least how much wrapping paper do you need? → look for the side area of the cylinder				
6					Seek the cylindrical side area	
7	Code "cylindrical side area"					
8		Calculate the area of the cylindrical side → calculate the area of the rectangle				
9					Calculate the area of the rectangle	
10			Area of the rectangle			
11		Seek the area of the rectangle → $2\pi r \times h$				
12					The length of the rectangle	

(continued)

Table 8.22 (continued)

	Visual	Production	Retrieval	Goal	Imagine	Manual
13		Seek the length of the rectangle → measure the length of the ruler				
14						Measure the length of the rectangle
15					The width of the rectangle	
16		Find the width of the rectangle → use a ruler to widen it				
17						Measure the width of the rectangle
18		Seek the area of the rectangle → length × width				
19						The area of the rectangle is length* wide
20						The area of the packing paper is the area of a rectangle, that is, the length and width of the paper
21						End

Table 8.24 presents a solution strategy in which each column represents the content of a module at different times in the problem-solving process. The leftmost column number represents the line number, and each row represents the cognitive logic step, which is not the same as the actual problem-solving process. The last line indicates the end of the cognitive process, that is, the end of the problem-solving process. RanA participated in the after-class oral report, as shown in Table 8.25. The left side of the table is the oral report data recorded during the experiment, and the right side is the cognitive process analysis of the oral report data, which was obtained via

Table 8.23 Analysis of the preclass oral report data of RanA on the "cylindrical flank area"

Oral report	Cognitive process analysis
[Read the title] A factory intends to manufacture a batch of chips (Fig. 8.7) that need to be packaged in the box on the side of the potato chip box. Can you help the factory calculate the number of wrapping papers needed for each box?	The text information is input through the reading list, and the propositional text frame and the question pattern are formed through visual coding
[Analysis] Use a piece of paper. Measure the paper to find the length and width and calculate the area by length × width (The primary test gives the student a chip can, an A4 sheet, ruler and scissors) (Student performs a hands-on operation) Take a paper box, and mark the position just on the perimeter. Use a straight line to cut off the extra part of the length; calculate the length and width of the paper (written on paper) 21 × 21.5 m 441.5 (square centimeter) Answer: 441.5 square centimeters are needed [Q: Can the wrapper wrap the chip box?] (Students performs hands-on operation) Paper wraps the box, finds that the paper is higher than the height. Marks the box height with a ruler line to indicate what will be part of the packaging paper and to cut off the excess height. The length and width of the paper are calculated: (written on paper) 21.5 × 14.2 = 305.3 (square centimeter) Answer: 441.5 square centimeters are needed The same: Say what you're thinking The first step is obtaining the paper packaging box. The second step is determining the box length and width. The third step is calculating the length × width to find the area for this piece of paper	(Inspired to understand, explains the problem-solving strategy, demonstrates clear thinking) (in accordance with the hands-on problem-solving strategies) (box length to meet the requirements, forgets that the paper height should be consistent and column side) Production rules: P1 Seeking cylindrical side area → Seeking the area of a rectangle, P2 Seek the area of a rectangle → length*width, P3 length*width, measure the length and width Manual: length, forgets the habit of examination and reflection in the process of solving the problem (after a simple prompt, encounters a difficulty with solving the problem) (Hands-on operation, student corrects the mistake by themselves) Manual: does not measure the length and width, writes the answer Correct calculation Retrieval: to change the length in program memory, the production type is automated to calculate the sum (Student's own summary)
[Diagnosis] RanA demonstrated clear problem-solving ideas and skilled operation but did not have the habit of checking the answer after the problem-solving process. The first time, she forgot the height of the chip can, and after being prompted to check after the discovery of the error, she corrected the problem-solving process	

Note the content of () is a part of the description

CMMPS analysis. The goal of the diagnosis is to divide the cognitive process on the right side of the analysis and compare it with the CMMPS-based problem-solving cognitive process (as shown in Table 8.24).

4. The first cognitive diagnosis intervention

Table 8.24 Analysis of the cognitive process of the "cylindrical side area" after class

	Visual	Production	Retrieval	Goal	Imagine	Manual
1	Visual coding					
2			Related semantic knowledge in long-term declarative memory			
3				The side area of it?	The side area of it?	
4	Code "the side area of it?"					
5		What is the side area of it? → which parts are included in the side area?				
6					Which parts are included in the side area?	
7	Code "cut one piece of cylinder side area"					
8		Side area → cylinder side area—cut area + two rectangular areas				
9					Cylinder side area—cut area + two rectangular areas	
10					Find the side area of the cylinder	
11			The side area of the cylinder			

(continued)

Table 8.24 (continued)

	Visual	Production	Retrieval	Goal	Imagine	Manual
12		Find the side area of the cylinder → 2πr*h				
13						The side area of the cylinder is 2πrh
14					Cut area	
15			Cut area			
16		How much to cut the area → cylindrical side area × 1/6				
17						The area cut is 2πrh × 1/6
18					2 rectangular areas	
19		Rectangular area → h × r				
20						The area of two rectangles is 2 × h × r
21		Side area → cylindrical side area—cut area + two rectangular areas				
22						2πrh—2πrh × 1/6 + 2 × h × r
23						End

The first cognitive diagnosis intervention was conducted December 21, 2011. RanA's first cognitive diagnostic intervention oral report analysis is shown in Table 8.26. The oral report data recorded during the experiment are shown on the left. On the right is the cognitive process analysis of the oral report data that was obtained via CMMPS analysis. The diagnostic result is based on the CMMPS-based problem-solving cognitive process (as shown in Table 8.24).

5. The second cognitive diagnosis intervention

The second cognitive intervention was conducted on December 22, 2011. The second cognitive diagnosis intervention oral report data for RanA is shown in Table 8.27. The left side of the form shows the oral report data recorded during the experiment; the right side shows the CMMPS analysis of the cognitive process.

Table 8.25 Analysis of after-class oral english report data of RanA classmate for "cylindrical side area"

Oral report	Cognitive process analysis
[Read] There is a cylinder, after cutting a piece, as shown in Fig. 8.11. Can you find its side area? Note that $h = 10$ cm, $r = 4$ cm	Enter the text information by reading the title, and perform visual coding to form a propositional text frame and question patterns
[Analysis] Use the area of the entire cylinder minus the area of the missing piece (Written on paper in 3 min and 7 s) $3.14 \times 4 \times 2 \times 10–4 \times 10$ $= 6.28 \times 2 \times 10–40$ $= 12.56 \times 10–40$ $= 125.6–40$ $= 85.6$ (square centimeters) A: The area is 85.6 square centimeters	Problem-solving strategy Goal: area of rectangle (Think: the missing part is rectangular, the area is $h \times r$) Manual: Think of "4" as "2" when performing the calculation (Solving steps are complete)
[Diagnosis] RanA made an error at the beginning of the problem-solving process. Namely, the cutoff part was mistaken for a rectangle and missed due to cutting off the addition of two rectangular areas. She did not understand the side with irregular graphics. The other error was due to a careless calculation process, which eventually led to an error in the solution	

Note () is a description

Table 8.26 Analysis of the first cognitive diagnosis of the "cylindrical side area" of RanA

Oral report	Cognitive process analysis
[Reading] There is a cylinder. After cutting one piece, as Tutu Fig. 8.11 shows, can you find its side area? $H = 10$ cm, $r = 4$ cm	The text information is input by reading the problem, and the propositional text frame and the problem pattern are formed through visual coding

(continued)

Table 8.26 (continued)

Oral report	Cognitive process analysis
Analysis (see their last topic, think for 1 min and 20 s) Q: How to solve the problem? The side area of the cylinder for this part, minus the missing part Question: Which side includes several parts? (by hand on the map to the surface [not included] to A) Ask: Think carefully Including the two parallelograms and the outside the area (left side cylindrical cut) Q: Why these two graphics? Because when you cut them out, it displays two faces. It includes two parallelograms, and outside the area (left side cylindrical cut) [Ask: Why there are two more graphics? Because of the cut piece, and the exposed surfaces becomes two pieces Question: What is the shape of these two graphics? Parallelogram [Q: How about you think for a minute?] Q: These two edges in the graph (mark the radius and height) form a vertical angle. What is their relationship? Vertical, right angle Q: What shape is it? Rectangle Q: Will you calculate the side area now? Yes (writes down the process of solving the problem in 3 min and 35 s) 3.14 * 2 * 4 * 12.56 * 2 * 10 = 12.56 * 2* 10 = 25.12 * 10 = 251.2 (cm) 4*10 * 2 = 40 * 2 = 80 (cm^2) 251.2 + 80 = 331.2 (cm^2) 4*10 = 40 (cm^2) 331.2 − 40 = 291.2 (cm^2) A: The area is 291.2 cm^2	Retrieval: activates declarative knowledge and procedural knowledge related to the side of the cylinder Goal: rectangular area. (In the previous problem-solving strategy, missed two sides of the rectangular area, does not understand the irregular graphics) Goal: two parallelograms + cut the remainder of the cylindrical side. (Thinks of the problem, thinks correctly, but mistakes the rectangle for a parallel quadrilateral, which looks like a parallelogram.) Retrieval: activates the related knowledge Retrieval: parallel quadrilateral knowledge (looks at a parallel quadrilateral) (a simple hint) Retrieval: activates the concept of right angles Production: Rectangular in the right angle and parallel quadrilateral Manual: calculates the whole side area of the cylinder Manual: calculates the area of two rectangles Manual: The area of the side of the cylinder plus the area of two rectangles. (Cuts off part of the area, mistakes for a rectangle minus the cutoff area)
[Diagnosis] Under gradually guidance, RanA came up with the correct problem-solving ideas but made some mistakes in cutting off the rectangular areas, leading to an error in problem solving	

Note () is a description

The diagnosis is obtained from the cognitive process analysis on the basis of the CMMPS problem-solving cognitive process (as shown in Table 8.24).

6. Comparative analysis of solving problems in different stages

Table 8.27 Analysis of the second cognitive diagnosis of the "cylindrical side area" of RanA

Oral report	Cognitive process analysis
[Read the question] There is a cylinder. After cutting a piece, as shown in Fig. 8.11, can you find the side area of it? $h = 10$ cm, $r = 4$ cm	Read the text information of the problem, and form the propositional text frame and the problem pattern after visual coding
[Analysis] Look at the question. What do you need to do? (Thinking for 10 s) [Q: How to cut the area of this face?] Multiply the radius by the height [Q: What is the surface cut off?] (Thinking for 15 s) [Q: Like this, what is the shape of the cut?] (Physical display side of the paper-packed cylinders) Round [Q: How to answer this question?] (Thinking for 30 s) Radius multiplied by high [Q: Is this rectangle cut off from a rectangular area?] No, it's round [Q: how to ask?] (Thinking for 48 s) [Q: What part is the cut area to the entire side?] The whole circle is 360 degrees, the cut off the part is 60 degrees, accounting for 60% [Q: how to figure out 60%?] (Thinking for 39 s) Accounting for one-sixth of the entire circle (side) $3.14 \times 4 \times 2 \times 10$ and then divide by 6 (Written on paper in 3 min and 30 s) $3.14 \times 4 \times 2 \times 10$ $= 12.56 \times 2 \times 10$ $= 25.12 \times 10$ $= 251.2$ (cm^2) $251.2 \div 6 \approx 41.9$ (cm^2) $251.2–41.9 = 209.3$ (cm^2) $4 \times 10 \times 2 = 40 \times 2 = 80$ (cm^2) $209.3 + 80 = 289.3$ (cm^2) A: Its side area is 289.3 (cm^2) [Q: Does this process tell you something about solving problems?] The side of the cylinder minus the missing piece, plus the inside of these two small rectangular areas; the missing piece of the entire cylinder side area is one-sixth of the whole	Retrieval: Knowledge of the side area of a cylinder Goal: The part that is cut is a rectangle. (The key reason for mistakes) (Guide the student to think about the shape of the cut surface) Visual: physical (Easy to answer correctly) Retrieval: fan area formula (Calculation error) Visual: cut shape (Shape to determine the correct answer) Retrieval: Activates the area calculation formula Retrieval: The degree of the circle and the known conditions (Correct understanding, but makes a calculation error, score knowledge is not strong) Rethinking (Self-reflection, obtains the correct answer) Manual: The area of the entire side of the cylinder Manual: Cut off part of the side area Manual: The area of the entire side minus the area of the side of the cutout Manual: The area of two rectangles are added by cutting Manual: Required side area (Summary of problem-solving ideas, clear thinking, correct calculation)
[Diagnosis] Summary: Through the physical display method, the student easily found that the cut surface was a cylindrical side, the side of the cut area accounted for one-sixth of the entire side, the problem-solving strategy was clear and definite, and the answer was calculated correctly	

Note () is a description

The characteristics of the problem-solving process of RanA at different stages are shown in Table 8.28.

As shown in the comparative analysis of Table 8.28, with a minimum number of prompts, RanA showed obvious changes in her problem-solving process, which helped her to answer the question correctly and form a good method for solving the problem.

7. Teacher interviews

With respect to the usual learning situation of RanA, her mathematics teacher, Han, reflected the following:

(1) RanA is a careful and hard-working student. After the completion of the task assigned by the teacher, she will read, preview, and find the previous topic to work on, but the result is not guaranteed to be correct.
(2) Of 90% of the questions that the teacher says can be solved, approximately half of the questions will be solved. The transformation of the question type will be prone to errors, and her ability to understand the problem independently is poor.

In the course of solving problems, RanA performed calculations carefully, and her ability to transfer side concepts was poor in the after-class inquiry question. After the intervention, the student could generate solution strategies and answer the questions correctly, and her performance was in line with that described by the classroom teacher.

Table 8.28 RanA's cognitive process changes in different stages

Stage	The characteristics of the problem-solving process
Preclass inquiry (December 17th)	The topic is relatively simple, the idea of solving the problem is clear, and the operation is smooth. However, there is no habit of checking the process of solving the problem, The height of the potato chip can was forgotten in the first operation. After examination, it was calculated, and the final problem was solved
After-class inquiry (December 20th)	It is difficult to explore the questions after class. In the beginning, the problem of solving the problem is wrong, and part of the cut is mistaken for a rectangle. However, the student missed two rectangular areas due to the new cut area. The side of the irregular figure is not understood. In addition, an error was made in the calculation process due to carelessness and ultimately led to a failure to solve the problem
First intervention (December 21st)	Pilot step-by-step guidance. Generate the right idea for solving the problem. The error that the cut part is rectangular led to a problem-solving error
Second intervention (December 22nd)	By showing the material object. The student easily finds that cut off surface is a cylindrical surface. The student figured out the side that had been cut off is the one-sixth of the whole side. Demonstrates clear thinking and correct calculation

8.3.4 Discussion

(一) Role of physical display or operations in problem solving

The "cylindrical flank area" tests the ability of students to apply what they have learned to solve practical problems. The area of the rectangle is 94% of the union, but many students do not calculate the area of the cylindrical side wrapping paper. The reason is that the cylindrical side wrapping paper is converted into a rectangular shape, increasing the difficulty. This difficulty also occurs in after-school inquiry question. The thinking of students in the stage of concrete operation is not formalized; thus, they require specific support. The physical display or operation can help the students perform this transformation smoothly. Thus, a complex problem can be transformed into a simple problem, which is helpful for solving the problem.

(二) Reviewing during the process of problem solving

With respect to the problem of solving the "cylindrical side area", some students correctly solved the problem, but because of carelessness, their calculation was incorrect. After inspection, they were able to quickly and accurately correct their error. Both Polya's and Hefield's mathematical problem-solving models underscore the importance of inspection and checking. Huntfield noted that inspection is a valuable activity. In a narrow sense, some trivial mistakes can be identified through inspection; in a broad sense, we can often find other ways to solve the problem, find connections with other problems, and sometimes from the process of solving problems, we can extract useful information for other situations and then become a better problem solver.

(三) Motivation and beliefs in the process of problem solving are highly important.

In solving the "trademark paper area" experiment, some students give up on thinking to answer the question in a very short period of time. However, the teacher prompted them to think carefully several times, and they eventually answered the question correctly. Jonathan's research shows that motivation and beliefs also play important roles in problem solving. ① Polygam additionally argues that "the extent to which you get involved in the question will depend on how ardent you are in solving it. Unless you have a very strong desire to do so, the possibility of solving a real problem is very small" (Georgia, 1987). The role of motivation and beliefs in the problem-solving process is a distinct area of study. However, it is not the focus of this study and so is only briefly discussed here.

8.4 Explanation of the Experimental Results for Mathematics Teaching

8.4.1 Starting from the Real Life, Creating a Problem Situation and Designing a Typical Problem

Many studies (Verschaffel, 2002) have shown that providing more opportunities for students to solve math problems in real-life situations helps them transform real problems into mathematical models and improves their mathematical problem-solving ability. With respect to the background of or problem situation, Jonathan reported that the problem situation generally needs to meet the following conditions: (1) include background knowledge of the story; (2) be from a specialized field; (3) have a time limit; (4) all the elements are interrelated; (5) have a generally acceptable solution; and (6) stimulate the willingness to solve the problem (Jonassen, 2000). In the process of mathematics teaching, teachers should create situations based on reality, avoid pseudo situations and pseudo applications, and avoid mathematical modelling activities at low levels and artisan operations. This study focused on the "mode" and "cylindrical flank area" and other areas of knowledge and presented a problem situation according to the reality of students' lives, the typical design problem, and the basis and principles of design.

8.4.2 The Problem-Solving Stage Is Integrated into Classroom Teaching to Help Students Form Good Ideas for Solving Problems

Polya's division of mathematical problem-solving stages provided a basis for cultivating students' problem-solving ideas. This finding suggests that teachers should not directly present the problem-solving process to students while they are lecturing but should adopt the method of "lecturing by doing" and integrate the problem-solving stages implicitly into the problem-solving process. For example, consider the unknown quantity first. What are the known data? What are the conditions? After a period of time, the students will also form a habit of solving the problem in this way and then form good ideas of how to solve the problem.

8.4.3 When a Student Has Difficulty Solving a Problem, the Teacher Provides an Appropriate Hint

Students will inevitably encounter difficulties in the process of solving problems, and they should be given enough time to think. Students should be trained to solve the problem and when encountering difficulties, to first think of the habit of problem solving. If students fail to make progress after a period of thinking, teachers should provide as few hints as possible and gradually guide students to find solutions to the problems and answer the questions correctly.

8.4.4 *Pay Attention to the Cultivation of Students' Interests, Attitudes, and Willpower*

The Schoenfeld survey revealed that students' incorrect attitudes and beliefs about mathematics are important factors influencing their problem-solving performance and that these incorrect attitudes and beliefs are formed through students' school experiences (Schoenfeld, 1985). Polya also stressed that education will teach students problem solving when students solve problems that are not too easy for them, when they learn that defeat is not hungry, when they learn to appreciate a little progress, when they wait for the main idea, and when they leave out what is not the main idea. If students have no chance to strive to solve a problem, then their mathematics education will fail (Polya, 1982). In school education, teachers should consciously cultivate these factors through a daily teaching plan, correctly apply these rules in class and ask questions. Through lectures, teachers should perform observations and practices for many years so that students gradually develop good problem-solving skills.

8.5 Summary

This chapter selects the typical knowledge points of declarative knowledge and procedural knowledge to apply the "one-on-one" cognitive diagnosis method to the practice of primary school mathematics teaching.

In terms of the "mode" knowledge diagnosis and intervention, 28 students participated in the oral report experiment, 18 students participated in the first cognitive diagnosis intervention, and 13 students participated in the second cognitive diagnostic intervention. Two cognitive diagnoses during the intervention for each student provided statistical data and strategy choices for two questions, resulting in a total of 118 passenger oral report experiments. We encoded and analyzed the oral report and combined all the statistics with the case in-depth analysis. The results show that the changes in the cognitive process during different stages of the "one-on-one" cognitive diagnosis intervention are significant and help student to solve the problem correctly.

In a sixth-grade class of 50 students, the topic of "the cylindrical lateral area of the knowledge point of diagnosis and intervention was the cylindrical side area" was used to construct a knowledge test. Forty-eight students participated in the preclass oral report experiment, 47 students participated in the after-class oral report experiment, 46 students participated in the first cognitive diagnosis intervention, and 5 students participated in the second cognitive diagnosis intervention. A total of 146 participants were in the oral report experiment. We encoded and analyzed the oral report and combined all the statistics with in-depth analysis. The results show that the changes in the cognitive process during different stages of the "one-on-one" cognitive diagnosis intervention are significant and helped students solve the problem correctly.

The prospect of "one-on-one" cognitive diagnosis is favored by many educators, and the diagnostic effect is remarkable. However, this method is not suitable for large portions of daily mathematics classroom teaching.

References

Anand, P. G., & Ross, S. M. (1987). Using computer-assisted instruction to personalize arithmetic materials for elementary school children. *Journal of Educational Psychology, 79*, 72–78.

Course Textbook Institute. (2006). *Compulsory education curriculum standard experimental textbook mathematics* (grade five). People's Education Press.

De Smedt, B., Grabner, R. H., & Studer, B. (2006). Oscillatory EEG correlates of arithmetic strategy use in addition and subtraction. *Experimental Brain Research, 195*, 635–642.

Fu, X. L., & He, H. D. (1995). A process of problem characterization. *Acta Psychologica Sinica.*

Georgia, G. (1987). *The discovery of mathematics, volume II* (Y. T. Liu, Trans.). Science Press.

Jonassen, D. H. (2000). Toward a design theory of problem solving. *Educational Technology Research and Development, 48*, 63–85.

Polya, G. (1982). *How to solve it* (Y. S. Yan, Trans.). Science Press.

Schoenfeld, A. H. (1985). *Mathematical problem solving*. Academic Press.

Shi, T. (1985). Cognition model of algebraic application problem. *Acta Psychologica Sinica.*

Verschaffel, L. (2002). *Everyday knowledge and mathematical modelling of school word problems.* Kluwer Academic.

Verschaffel, L., & De Corte, E. (1997). Teaching realistic mathematical modeling in the elementary school: A teaching experiment with fifth graders. *Journal for Research in Mathematics Education, 28*, 577–601.

Chapter 9
Cognitive Simulation of the Interaction Process in a Math Class

More than 2000 years ago, the "heuristic" of Confucius (551–47 BC) and the "maternity witchcraft" of Socrates (470–39 BC) were major questioning methods in education. By asking questions, educators guide students to think and ultimately achieve their learning goals. According to the interaction between teachers and students by Confucius and Socrates, the content of questions should be arranged in an orderly, logical and cognitive manner.

Currently, classroom instruction is still the major format of school education. Classroom interaction is important in classroom teaching. Classroom instruction is a practicum of cognitive activities (Zhong, 2012). Recent studies have shown that interactions between teachers and students in classrooms essentially involve a series of social cognitive processes (Schwarz et al., 2009; Wedin, 2010). This book studies existing methods for analyzing classroom interactions from the perspectives of the behavioral system and the information system. Furthermore, this book analyzes classroom interactions as learning processes on the basis of brain science, cognitive neuroscience, psychology, and artificial intelligence. Finally, this book simulates the interactions in ACT-Rim to better understand the cognitive processes and to help teachers understand the learning process thoroughly and thus design effective teaching methods to assist students in learning.

9.1 Existing Classroom Interaction Analysis Method

9.1.1 Behavioral System Perspective

From the perspective of the behavior system, studies of teaching focus on two aspects. The first is the teaching value of studying specific behaviors.

Studies that have focused on the relationship between learning persistence and the teaching interaction of adult learners have shown that 26% of learners agree that

X. Wei, *Artificial Intelligence in Mathematics Education*, Smart Computing and Intelligence, https://doi.org/10.1007/978-981-95-1405-2_9

asynchronous discussion is positively related to persistent learning. Wai King Tsang's study examined the interaction between teachers' feedback in nonnative English classes and students' hand-raised oral reports. The results revealed the following: (1) Redoing may trigger other forms of feedback. (2) Although redoing and explicit modification help correct spelling mistakes, discussion, and consultations are more conducive to the correction of grammatical errors (Tsang, 2004). Researchers such as Judith Kleine have examined the relationships between different formats of interaction (such as face-to-face communication without using a computer, computer-based collaborative asynchronous communication, and computer-assisted face-to-face communication). Research has shown that computer-mediated interactions are more regulated; however, there is more learning in face-to-face interactions than in computer-mediated interactions.

Another emphasis of research on pedagogical analysis conducted from the behavioral system perspective is a detailed hermeneutic analysis of interactions. The topics include an effective and reliable method for defining the structure and characteristics of the dialog, the role of dialog as understood through interactive dialog analysis, and the computing model of dialog in the intelligent educational system (Pilkington, 2001). Better interactive quality results in a higher level of interaction (Moore & Marra, 2008).

Typical classroom interaction analysis methods include the Flanders interaction analysis system (FIAS) and student–teacher analysis. The Flanders interaction analysis system is a classroom behavior analysis technology proposed by Flanders while at the University of Minnesota in the 1960s. It is used to record and analyze the processes and impact of classroom lingual interactions between teachers and students. The system consists of three main parts: (1) a set of coding systems for describing interactive behaviors in classrooms; (2) a set of standards for observing and recording coding standards; and (3) a matrix for displaying data for analysis. Ning Hong et al. used the Flanders interaction analysis system to analyze a middle school physics class and ameliorate the shortcomings of FIAS (Ning & Wu, 2003). S-T analysis is used mainly for quantitative analysis of classroom interactions. S-T analysis of teaching behaviors is divided into S (student) behaviors and T (teacher) behaviors. It moreover divides teaching into four different teaching modes: the practice mode, the lecture mode, the conversational mode, and the mixed mode (Fu & Zhang, 2001). The S-T analysis results can be represented with S-T charts. Shan Yingjie from Shaanxi Normal University used S-T analysis of educational technology to analyze education processes in the Educational Technology Research Methods course, the TV Principles course, and six other specialized courses (Shan, 2008).

9.1.2 Information System Perspective

In the field of teaching analysis, some scholars regard teaching as a process of information flow. Professor Li Kedong carried out information flow analysis on the cognitive learning process in the teaching system and adopted functional simulation, a method of systematic scientific research, to analyze the teaching system (Li, 1990). A. Dean Hauenstein introduced the concept of systems into instruction and made it clear that all systems are cyclic processes of input, process, output, and feedback. He suggested that the instruction system was an information system (Malan & Sheng, 2005).

Professor Yang Kaicheng also considers the instruction system essentially an information system. The analysis of the teaching system is actually an analysis operation, which uses another coding system to characterize the natural language representation of the teaching system, builds a teaching analysis from the perspective of information systems, starts from the overall functional mechanism of the teaching system, and introduces the teaching analysis from the perspective of the information system into the actual analysis of the operation (Yang, 2007). He proposed IIS (instructional information set) graph analysis. Finally, Lin et al.'s empirical research proved that the activation of target knowledge and learning outcomes are positively related (Lin, 2009).

9.1.3 Comments on Existing Analytical Methods

(1) The Flanders interaction analysis system (FIAS) mostly uses fixed time units (such as every 3 s) to collect data, which can easily lead to "meaning unit" segmentation. Additionally, the observation scale quantifies only the language behaviors of teachers and students, which is too approximate to capture the whole-class interaction process and thus cannot reflect all the interactions in a class. It pays more attention to teachers' (or students') language behaviors, such as the proportions of teachers' and students' speech, the number of teachers' questions, and the number of students' answers. The analysis is moreover limited to the level of explicit behavior.

(2) In S–T analysis, the behavior definitions of S and T are inexplicit. Through the S–T analysis chart, we can determine the amount, proportion and time of the teachers' and students' behaviors, but we cannot identify how they have behaved. Additionally, we can identify the teaching mode, but the evaluation of the teaching process is vague.

(3) IIS graph analysis is an improvement of the former methods that is based on the analysis of behavioral systems. It pays more attention to the content of teaching and predicts teaching effectiveness by analyzing the activation of knowledge in the process of classroom interaction. However, IIS graph analysis focuses on the input and output of teachers and students, considering that internal information processing is transparent and invisible; that is, it does not consider the internal information processing of the students.

9.2 Classroom Interaction Cognition Analysis and Simulation

9.2.1 The Boom of the Learning Sciences

Currently, many countries strongly support research in brain science and the learning sciences. Arden Bement, a chief executive at the National Science Foundation, asserted: "Fundamental research on learning is important. In today's complex and rapidly changing environment, a basic understanding of the learning process is to help us to develop a knowledge base that is necessary for the prosperity of a world that is forever in flux." Beijing Normal University Cognitive Neuroscience and Learning State Key Laboratory and Southeast University Learning Science Research Center are conducting research in this area.

The development of the learning sciences provides a new perspective for effective research and study. The learning sciences constitute an interdisciplinary field, as the following quotes demonstrate: "It incorporates a variety of theoretical perspectives and research paradigms about human science in order to understand the nature and conditions of learning, cognition, and development." "The goal of learning science is primarily to better understand cognitive processes and social processes to produce the most effective learning and, secondly, to redesign our classrooms and other learning environments with the knowledge of science so that Learners can learn more effectively and deeply (Sawyer, 2006)." The learning sciences advocate that learning be placed in a broad perspective of multidisciplinary research, which covers many fields of research, such as information science, brain science, cognitive science, psychology, and education, and applies the latest developments in brain science to the learning and education process by building bridges between the mind, the brain, and education.

The U.S. Department of Education's Office of Educational Technology released the National Educational Technology Plan 2010 ("Plan") on March 5, 2010, entitled "Changing American Education: Technology to Make Learn More Powerful." The term "learning science" appears many times in the plan. The latest research in the learning sciences reveals the process of how people learn, which provides an important theoretical basis for the application of educational technology. The plan also acknowledges that current education systems focus little attention to students' thinking processes, place too much emphasis on the mastery of factual knowledge when assessing students, and do not focus on students' feedback and improvement of immediate learning during the learning process. This situation is also common in the Chinese education system. The continuous development of the learning sciences provides an important foundation to study classroom interactions from the perspective of the learning process.

9.2.2 Classroom Interaction Cognitive Analysis Framework

According to the current psychology and cognitive neuroscience research results (Banks, 2009; Nader & Hardt, 2009; Torey, 2009; Wilson et al., 2004), the classroom interaction cognitive analysis framework—CAUT (cognitive architecture of human thinking) (Cui et al., 2011)—is shown in Fig. 9.1.

CAUT pays close attention to students' thinking processes and particularly to the understanding of the learning process. The model includes the following parts: sensory organs, sensory controllers, object perception, long-term declarative memory, long-term procedural memory, active objects (working memory), control and decision-making, situation-goal-anticipation, motor control, device effects, the

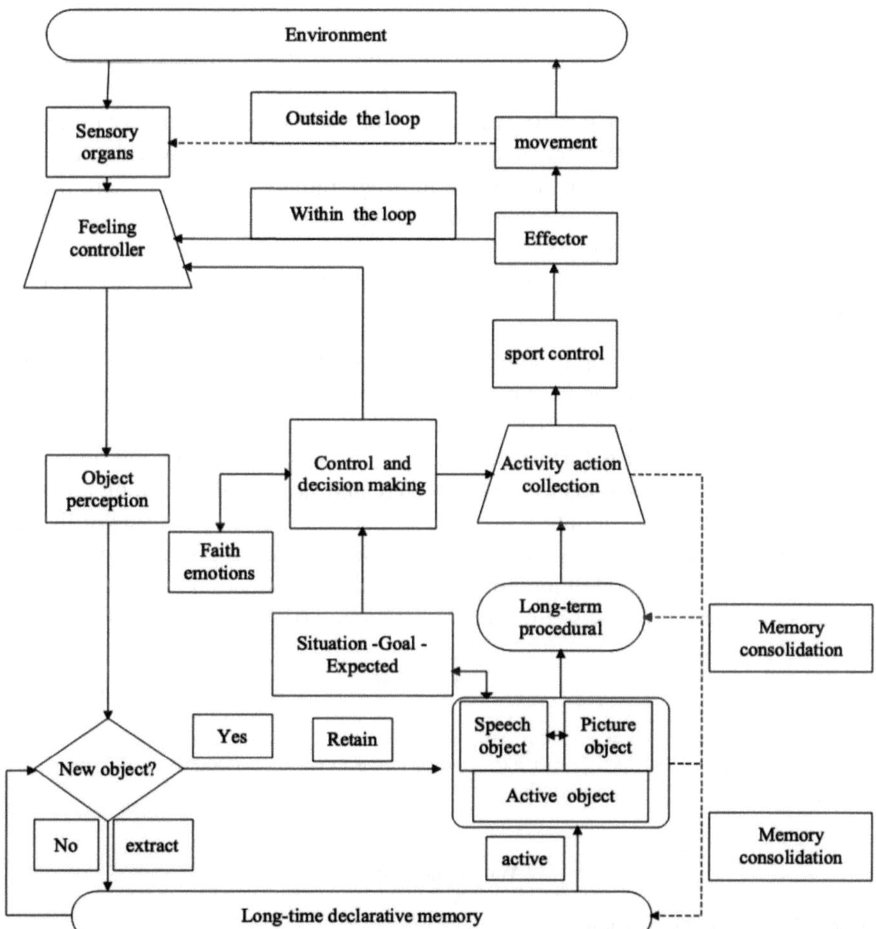

Fig. 9.1 CAUT

outer loop, and the inner loop. To further describe CAUT, it can be divided into the following eight modules: (1) External loop/Internal loop: internal loop or external loop; (2) IO (internal object): internal object; MC (declarative memory retrieval check): declarative memory extraction to determine whether the perception of the object reveals a new object or the old object; (4) LTDMO (long-term declarative memory operation): long-term declarative memory cognitive operation; (5) AO (active object buffer): activation of the collection of objects, including seeing and hearing the object, is a part of working memory (working memory); (6) AADM (active action buffer and decision-making): activation of action, decision-making and related parts; (7) CGE (context, goal, expectation): context, target, and expectation related to current task; and (8) Action: action module, including motion control and effectors. The sequence of modules in CAUT is shown as in Fig. 9.2.

When we perceive (see or hear) an external object, we consider it an internal object by visual or auditory channel coding and then determine if the internal object is stored in the object and activate the corresponding object if it has already been stored in long-term declarative memory; otherwise, it will be repeated and directly enter into the active object (part of the working memory) (for example, in real life, we remember that a strange phone number needs to be repeated). The objects in working memory activate corresponding actions in long-term programmatic memory. There may be more than one active action, and one action is selected through decision-making.

Compared with other cognitive structures such as ACT-R (Anderson et al., 2004), SOAR (Laird et al., 1987), and CLARION (Naveh & Sun, 2004), the model has the following features:

(1) An internal speech loop is added between the effector and the sensory organs, and the existence of the loop has been demonstrated in cognitive neuroscience (Pulvermüller & Fadiga, 2010; Wiley, 2006).
(2) Long-term memory is further divided into declarative memory and procedural memory and corresponds to the knowledge and skills in the learning process.
(3) Emphasize the consolidation of memory. Recent studies have shown that the use of long-term memory in learning or other cognitive processes is separate from the consolidation of memory and that the consolidation of memory occurs after cognitive processes (Born & Diekelmann, 2010; Maquet, 2001).

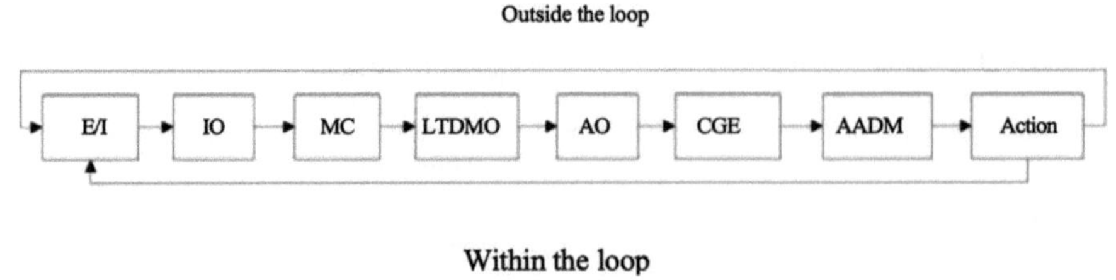

Fig. 9.2 Modular representation

9.2.3 Cognitive Analysis and Simulation of Classroom Interaction

(一) Research subjects

We selected a math class from the seventh grade (second semester). The content is from Chapter I, "Rich graphics world," Section 4, "Looking from different directions." The textbook is a compulsory education curriculum standards experimental textbook published by the Beijing Normal University Publishing Group for grade seven mathematics (May 2005 fourth edition). The instructor is Tang Lujun from Jinan Yuying Middle School.

(二) Typical classroom interactive sequence

For research purposes, the class video was converted into text. While watching the video, we found that teachers often used in-kind (or teaching aids) and multimedia courseware to help students understand why they were in a middle school mathematics class. Therefore, not only did the teacher's classroom discourse need to be converted into text during the conversion process, but the entity displayed and the contents on the big screen were also recorded using annotation and the addition of remarks.

In this study, the "teaching goal-teaching subgoal" approach classified class interactive text. The teaching objectives were established based on Bloom's educational taxonomy of objectives (revised version) (Aderson, 2009) in terms of knowledge and cognitive process analysis. We classified the teaching activities in the interactive texts of "Looking from Different Directions" and obtained 9 types of teaching activities in time sequences. We chose an interactive sequence to explain the concept, as shown in Fig. 9.3.

(三) Interactive sequence cognitive process simulation

ACT-R has been a well-known cognitive simulation tool for many years at the Cognitive Science Laboratory led by Anderson, a famous cognitive psychologist at Carnegie Mellon University in the United States. Its internal structure and parameter settings are based on a large amount of cognitive psychology experimental data. Most of the data have been verified by NMR experiments. The extrinsic format is a programming language. The programs written in this programming language correspond to the cognitive preconditions of the ACT-R, which is consistent with the cognitive process of real-life experiments and can achieve simulation results. It has been widely used to simulate different aspects of human cognitive behavior, such as the hornet problem, language comprehension, pattern recognition, memory, and simple geometric proofs.

The simulation of the learning process is very complicated and needs to be analyzed in a particular context for each sentence. Figure 9.3 shows a typical classroom interaction sequence. Owing to space limitations, this section selects only the

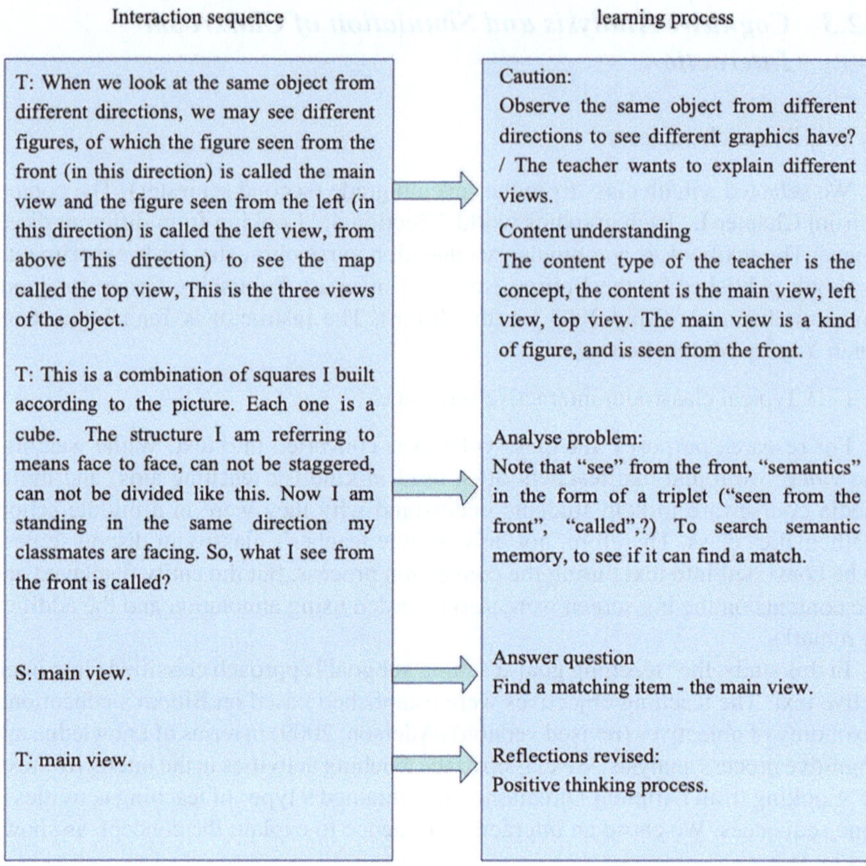

Fig. 9.3 Explaining the concept of the "main view" in the classroom interaction sequence

classroom interaction sequence of "teacher question–student answer" for simulation analysis to provide a method for classroom interaction analysis. The interaction sequence is shown in Fig. 9.3, where T is the teacher and S is the student.

T: "Well, what I see from the front is called…?"
S: "Main view."

The CAUT model is used to analyze the interactive sequence of the selected class. The results of the analysis are transformed into a program (Lisp programming language) that can be executed in ACT-R to simulate the learning process.

Before asking questions, the teacher has already discussed the concept of "the main view," which assumes that the student's long-term declarative memory stores "the front view as the main view." The following describes the teacher questioning and the cognitive process of student responses:

(1) Students hear the teacher's words: "Well, what I see from the front is called…?". The words enter the auditory pathway in the brain (such as the vestibular pathway) and undergo neural coding; (2) the encoded word activation-related objects are triggered in the mental lexicon of LTDMO and into the active object set (AO); (3) the content of the set of active objects (part of the working memory) is semantically understood, and the target of the sentence is set to search for the problem of searching for triples (seen from the front as?); (4) the active object activates a production rule in long-term procedural memory and generates the corresponding action; (5) there may be more than one active action, and one of the actions is selected through the "decision"; (6) the form (viewed from the front as?) is searched in long-term procedural memory; (7) there is only one with the answer "main view," and the search ends; and (8) the students say the answer.

To visualize what students hear, "So, now I see from the front is called?", answer the "main view" of the cognitive process, with M rows and eight columns of the cognitive matrix to represent, as shown in Fig. 9.4. The numbers on the left indicate the line numbers, each line represents the cognitive logic step, not the actual step, and the last line indicates the end of cognition. The eight columns correspond to the eight modules in Fig. 9.2. As shown in Fig. 9.4, the target is set in line 7 until the student understands the teacher in line 9. Line 12 gives the answer, that is, the student reaches their goal, and the cognitive task ends.

ACT-R provides an abstract cognitive structure that describes the cognitive model from a functional point of view only. In this study, we need to write a program that can be simulated in ACT-R according to the analysis process of the cognitive matrix. The program is written in the Lisp programming language. The simulation results are shown in Fig. 9.5. The minimum time interval is 0.55 s.

Through the simulation, we find that, in answering the questions, students must first determine the goal and highlight its importance. This sequence is in line with the teaching task of "informing learners' goals" in the "Nine Teaching Events" proposed by Gagne. Based on the objects in the activity set, match generation is activated (i.e., production begins), and extraction is started in long-term declarative memory (start retrieval); when multiple productions are activated at the same time, conflict resolution is adopted, and one of the productions is executed. Specifically, at 0.5 s, one of the productions is activated to search for long-term declarative memory to find the matching content. The goal is achieved, and the cognitive process is complete.

(四) Analysis of the simulation results

Through the simulation mentioned above, the internal cognitive process of students' answering questions is deeply examined. This internal cognitive process is demonstrated by ACT-R. The declarative knowledge and procedural knowledge involved in this learning activity are extracted. Depending on the type of knowledge, teachers use different teaching methods, such as the main view, left view, proposition and other declarative knowledge, mainly provided by the teacher, to explain memory. For example, arithmetic computing, solving equations, and geometric proofs of procedural knowledge require students to engage in the actual training process.

	E/I	IO	MC	LTDMO	AO	CGE	AADM	Action
1	E	Then	repeat	activation	Then			
2	E	Me	repeat	activation	Then me			
3	E	From	repeat	activation	Then I'm from			
4	E	positive	repeat	activation	Then I am from the front			
5	E	see	repeat	activation	Then I see it fron the front			
6	E	of	repeat	activation	Then I see it fron the front			
7	E	Called?	repeat	activation	So what do I see fron the front?	Goal:Search the problem		
8	E		repeat	activation	Sowhat do I see fron the front?	extract		
9	E		repeat	activation	Sowhat do I see fron the front?	(Seen from the front called?)	Seen from the front called?	Seen from the front called?
10	I	Seen from the front called?	repeat	Seen from the front called?	Seen from the front called the main view	Answer: main view	Seen from the front called the main view	Seen from the front called the main view
11	I				Seen from the front called the main view	The goal is consistent with the answer	Seen from the front called the main view	Seen from the front called the main view
12						finish		

Fig. 9.4 Cognitive matrix description of the student answer to the "main view" problem

The model can also analyze whether teacher–student interactions can effectively promote students' cognition and meet students' cognitive rules. For example, two different ways of asking questions, "Do you see from the front?" and "Do you see the front view from the front?", fall into two different categories: the former is a search problem, and the latter is a judgment problem. The cognitive processes of answering these two types of questions in ACT-R differ, and students exhibit different levels of procedural knowledge in answering these two types of questions.

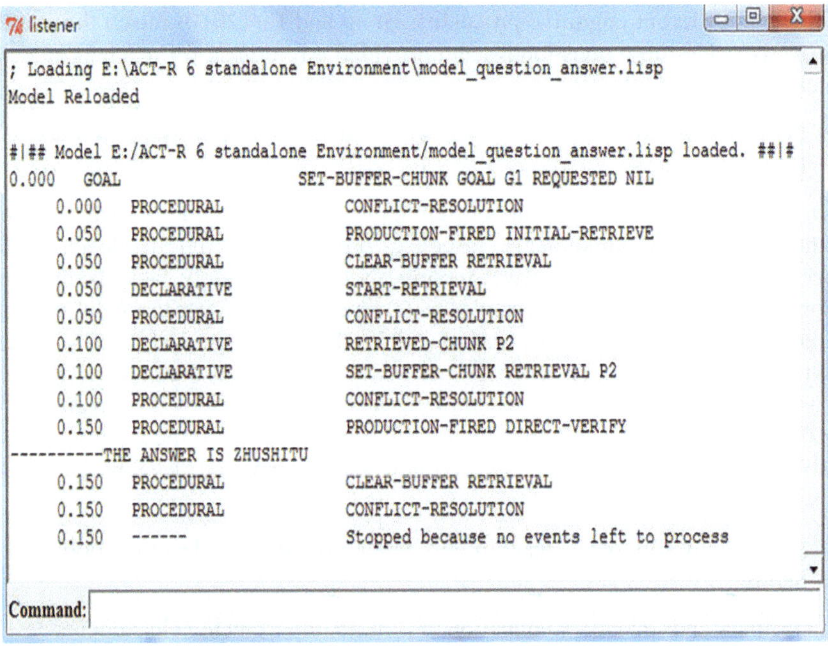

Fig. 9.5 Simulation of cognitive processes

9.3 Explanation of Cognitive Process Analysis of Classroom Instruction

Through the analysis above, we find that different classroom interactions produce different learning processes, which will lead to different learning outcomes. Therefore, teachers should pay attention to the following three aspects of classroom teaching:

(1) Carefully designed classroom questions promote students' deep understanding

In primary and secondary school classrooms, asking questions is still a common method of classroom interaction. However, from actual classroom observations, teachers ask simple, casual, or even repeated questions to enliven the classroom, some questions lack a scientific basis and scientific design, low-level questions are posed, and there is excessive inhibition of pure memory problems in the development of students' thinking, none of which is conducive to a deep understanding of the teaching content. Teachers should carefully design effective classroom questions based on the characteristics of students' cognition and content so that students can automatically establish connections between old and new knowledge and deepen their understanding of the learning content when answering questions. As seen from the simulation process above, different questions were asked about the "main view," such as "What is seen as a positive?" and "Is the main view seen from the front?", resulting in different cognitive processes. Hong and Lu (2010) raised the standard of effective classroom questioning and provided a reference for teachers to design effective classroom questions.

(2) Providing reasonable feedback enables students to actively participate in the learning process

In classroom teaching, teachers often apply simple evaluations such as "good," "right," and "wrong" to students' responses. Providing a single feedback method, especially for students who answer incorrectly or incompletely, lacks further inspiration and induction. It is impossible for all students to answer teachers' questions correctly. Thus, reasonable inspiration and feedback are essential; even if students provide the correct answer, teachers can ask students about the process of finding the answer, such as "How did you get to this answer?" and "Why do you answer like this?". Such questions can help students pay more attention to the learning process, establish connections between old and new knowledge, and develop the habit of knowing what they are doing.

(3) Scientific design of the teaching process helps students develop good thinking habits

In primary and secondary education, it is more important to help students develop good thinking habits than help them simply acquire knowledge. Good habits of thinking can help students engage in smooth knowledge transfer when they encounter similar problems or new problems and even creative problem solving. Thinking habits are an important part of tacit knowledge and an important part of procedural memory. Their development is a long-term process. The development of thinking habits should be related to the teaching of specific subject knowledge, which requires teachers to explain the typical problems of knowledge point design during instruction. Each step of the explanation process should consciously train students' thinking ability and place a greater emphasis on the problem-solving process so that students continue to develop good thinking habits when problem solving. For example, when the concept of "modalities" in the fifth grade textbook (second volume) is explained, it is common practice to provide questions and data and then tell the students that "the most frequent occurrence of a set of data is the mode of the set of data." Another method is to state the reality, such as "The school has agreed to hold a birthday celebration next year for grade five, class three. However, only the birthdays of students born in a certain month can be celebrated. If you were the class teacher, how would you choose the month? Which month do you think should be chosen?". This example is close to the students' lives; students can therefore collect data and select the month in which the most students who have a birthday according to the statistical results and thereby grasp the concept of the "mode." Although two different teaching processes allow students to learn the "plural" concept, there is a difference in the degree of conceptual understanding and ability to use the "plural" concept to solve practical problems. During classroom teaching, teachers should help students develop thinking habits to use mathematical knowledge to solve practical problems.

9.4 Summary

This chapter analyzes the existing methods of classroom interaction analysis from the perspectives of the behavior system and information system and, on this basis, proposes analyzing classroom interaction from the perspective of the cognitive process. In accordance with the research results of brain science, cognitive psychology and cognitive neuroscience, this chapter proposes a classroom cognitive framework—i.e., CAUT. The typical classroom interaction sequence of "Reading from Different Directions" in a middle school (Grade 7) mathematics classroom was subsequently analyzed by using CAUT, and a cognitive simulation was implemented with ACT-R. According to the analysis and simulation results of classroom interaction cognition, three suggestions for classroom teaching are proposed to help teachers design more effective teaching methods.

However, during the process of conducting classroom interaction analysis based on the learning process, this chapter performs only a brief analysis. The preparation of a cognitive program requires a specific and profound understanding of the learning process. How to use the CAUT model to automatically generate cognitive programs and ACT-R to achieve whole-class simulation needs to be further studied.

References

Aderson, L. W. (Ed.). (2009). *Bloom's educational objective taxonomy* (Rev ed., X. P. Jiang, Trans.). Foreign Language Teaching and Research Press.

Anderson, J. R., Bothell, D., Byrne, M. D., Douglass, S., Lebiere, C., & Qin, Y. (2004). An integrated theory of the mind. *Psychological Review, 111*, 1036–1060.

Jerome, H., & Freiberg, K. (1981). Three decades of the Landers Interaction Analysis System. *Journal of Classroom Interaction, 16*, 1–7.

Banks, W. P. (2009). *Encyclopedia of consciousness* (pp. 389–402). Elsevier.

Born, J., & Diekelmann, S. (2010). The memory function of sleep. *Nature Reviews Neuroscience, 11*, 114–126.

Cui, G., Wei, X., & Boling, L. W. (2011). A cognitive model of human thinking. In *Proceedings of the Seventh International Conference on Natural Computation (ICNC)* (Vol. 2, pp. 31–37). IEEE.

Fu, D. R., & Zhang, H. M. (2001). *Educational information processing* (pp. 94–108). Beijing Normal University Press.

Hong, S. Z., & Lu, Z. Z. (2010). Effective classroom questioning teachers: The value orientation and standard construction. *Education Research, 31*(4), 65–70.

Jerome, H., & Freiberg, K. (1981). Three decades of the landers interaction analysis system. *Journal of Classroom Interaction, 16*, 1–7.

Laird, J. E., Newell, A., & Rosenbloom, P. S. (1987). Soar: An architecture for general intelligence. *Artificial Intelligence, 33*, 1–64.

Li, K. D. (1990). The application of systematic science method in teaching system research. *Teacher Education Research*, 4.

Lin, F. (2009). *Teaching system analysis IIS map method* (pp. 47–52) (Master's degree thesis). Beijing Normal University.

Malan, & Sheng, Q. L. (2005). A new framework of educational objectives classification-review of the integration of Hornstein teaching system and target classification. *China Electrochemical Education*, 7.
Maquet, P. (2001). The role of sleep in learning and memory. *Science, 294*, 1048–1052.
McGettigan, C., Eisner, F., & Scott, S. K. (2009). A little more conversation, a little less action—Condition of roles for the motor cortex in speech perception. *Nature Reviews Neuroscience, 10*, 295–302.
Moore, J. L., & Marra, R. M. (2008). A comparative analysis of online discussion participation provisions. *Journal of Research on Technology in Education, 38*, 191–212.
Nader, K., & Hardt, O. (2009). A single standard for memory: The case for reconsolidation. *Nature Reviews Neuroscience, 10*, 224–234.
Naveh, I., & Sun, R. (2004). Simulating organizational decision making with a cognitive architecture CLARION. *Journal of Artificial Society and Social Simulation, 7*, 1–39.
Ning, H., & Wu, J. H. (2003). To establish the relationship between quantitative structure and meaning of understanding-Flanders interactive analysis of the improved use of technology. *Education Research*, 5.
Pilkington, R. (2001). Analyzing educational dialogue interaction: Towards models that support learning. *International Journal of Artificial Intelligence in Education, 12*, 1–7.
Pulvermüller, F., & Fadiga, L. (2010). Active perception: Sensorimotor circuits as a cortical basis for language. *Nature Reviews Neuroscience, 11*, 351–360.
Sawyer, R. K. (2006). *The Cambridge handbook of the learning sciences*. Cambridge University Press.
Schwarz, B., Dreyfus, T., & Hershkowitz, R. (2009). *Transformation knowledge based classroom interaction* (pp. 58–71). Routledge.
Shan, Y. J. (2008). Analysis of the use of S-T analysis of educational technology classroom teaching problems. *Modern Education Technology*, 10.
Tello, S. F. (2007). An analysis of student persistence in online education. *International Journal of Information and Communication Technology Education, 3*, 47–60.
Torey, Z. (2009). *The crucible of consciousness: An integrated theory of mind and brain*. MIT Press.
Tsang, W. K. (2004). Feedback and update in teach-student interaction: An analysis of 18 English lessons in Hong Kong secondary classrooms. *RELC Journal, 35*, 187–209.
Vandern, H., Kleine, J., Krol, K., Peer, M., & Staarman, J. K. (2005). Interaction in three collaborative learning conditions. *The Journal of Classroom Interaction, 4*, 29–39.
Wedin, A. (2010). Classroom interaction: Potential or problem? The case of Kara we. *International Journal of Educational Development, 30*, 145–150.
Wiley, N. (2006). Inner speech as a language: A Saussurean inquiry. *Journal for the Theory of Social Behavior, 36*, 319–341.
Wilson, S. M., Saygin, A. P., Sereno, M. I., & Iacoboni, M. (2004). Listening to speech activates motor areas involved in speech production. *Nature Neuroscience, 7*, 701–702.
Yang, K. C. (2007). A preliminary study of teaching system analysis technology. *China Electrochemical Education*, 8.
Zhong, Q. Q. (2012). Classroom interaction research: Implications and issues. *Education Research*, 10.

Chapter 10
Cognitive Simulation of Problem-Solving Strategies in Mathematics for Elementary School Students with Learning Disabilities and Its Pedagogical Implications

The Compulsory Education Mathematics Curriculum Standards (2022 edition) emphasizes that the mathematics curriculum should pay attention to the teaching level, stimulate the learning interest of students with learning difficulties, encourage them to think positively, cultivate good learning habits, and adapt to the developmental needs of students (Ministry of Education of the People's Republic of China, 2022). Students with learning disabilities constitute a particular group to which front-line teachers give attention. This book proposes a cognitive simulation of problem-solving strategies based on existing research. Taking the problem of "plurality" in primary school mathematics as an example, this book, which is based on the cognitive problem-solving model in primary school mathematics, analyses the cognitive process of problem-solving strategies, writes a cognitive process, and performs a cognitive simulation. Afterward, students with learning disabilities were selected and interviewed by their teachers to understand their usual learning situation. The cognitive simulation of problem-solving strategies for students with learning disabilities helped visualize the internal problem-solving process, helped teachers develop personalized intervention schemes, helped students overcome bad problem-solving habits and develop good problem-solving strategies. Problem-solving strategies and cognitive simulation are essential for designing and developing intelligent tutoring systems, building innovative learning environments, and providing targeted cognitive diagnosis and intervention.

X. Wei, *Artificial Intelligence in Mathematics Education*, Smart Computing and Intelligence, https://doi.org/10.1007/978-981-95-1405-2_10

10.1 Introduction

10.1.1 Analysis of Students with Learning Disabilities

Learning disabilities (LDs) were first proposed by American scholar S. Kirk in 1962 and mainly refer to students with average intelligence whose academic performance is lagging (Bateman & Kirk, 1962). Kirk and Chalfant (1984) classified learning disabilities as developmental learning disabilities. Tournaki (2003) discussed the importance of strategy instruction and teaching on mathematical ability (MD) and the importance of direct tactics in addition to instruction for mathematical ability (MD). In recent years, many studies have focused on the characteristics of problem-solving strategies and the development of difficulties in primary school mathematics (addition and subtraction over 20 years) (Geary et al., 1999).

Research on learning difficulties of students with learning disabilities has been conducted using different methods. Bai et al. (2020) compared and explored the strategies used by students with learning difficulties and ordinary students in the process of solving subtraction problems and noted that students with learning difficulties chose borrowing strategies more often, whereas ordinary students chose decomposition strategies more often. Additionally, students with numerical difficulties demonstrated poor strategy implementation effectiveness and relatively rigid choices. Liu and Mao (2021) adopted a meta-analysis method to explore the effectiveness of interventions in improving the mathematics performance of students with mathematics learning difficulties and its regulatory factors. Yang et al. adopted a three-factor mixed experimental design and considered the cognitive processing of students with learning difficulties to analyze the influence of different levels of central executive load on the use of estimation strategies by students with math learning difficulties and ordinary students. Wang et al. explored the influence mechanism of working memory components on word problem solving in children with math learning difficulties and analyzed groups of students with learning difficulties, ordinary students, and excellent students from the aspects of the central executive system, visual template, and speech loop. Liu (2018) tested the word problem-solving ability of students in the fourth grade of primary school. The results revealed that students with math learning difficulties were able to identify information less effectively than ordinary students were and that the use of schema representation strategies could better help students with learning difficulties solve word problems. Zhang et al. used eye movement technology to investigate the influence of picture information on the ability of students to solve different subtypes of math problems and showed that picture information can promote the schematic representation of problems by students with math learning difficulties, thereby improving their problem-solving performance. Liu et al. adopted an experimental research method to investigate the types of visual representations of students with math learning difficulties in the third grade of primary school in solving word problems and their impact on word problem solving and further explored the role of examples in promoting the beneficial effect of visual representations in solving word problems on students with math learning difficulties.

10.1.2 Mathematical Problem-Solving Cognitive Simulation

The cognitive simulation of problem solving is a topic of interest to researchers in the learning sciences. To perform a cognitive simulation of problem solving in junior high school mathematics, Professor Anderson et al. (2008) of the School of Psychology, Carnegie Mellon University in the U.S., used ACT-R (adaptive control of thought-rational) to implement a cognitive simulation of the process of solving the algebraic equation "$7x + 3 = 38$." This simulation was performed using ACT-R (adaptive control of thought-rational Anderson et al., 2008). Cui et al. used the ACT-R 7.0 software developed by Anderson's team and the Pyactr package developed by Adrian Brasoveanu et al., which improved language compatibility, and applied the cognitive simulation method to explore the cognitive processes and differences between the profit and loss model and the absolute value model of rational number addition. In terms of the cognitive simulation of problem solving in primary school mathematics, Wei et al. implemented a cognitive simulation of problem solving in primary school mathematics to map the brain areas activated at a particular moment of problem solving and the brain's blood oxygen-level dependent response (blood oxygen level). The corresponding data of the brain's blood oxygen level-dependent response (BOLD) are presented (Cui & Wei, 2013; Wei et al., 2012). Based on the cognitive process analysis method, Zhang et al. conducted cognitive process analysis and cognitive simulation for the process of solving the third-grade mathematics "comparison" problem and compared the cognitive output and extraction path of teachers and students in the cognitive process. Regarding cognitive simulation of geometry proof problems, Gelernter et al. (1960) developed a computer program, Geometry Machine, to simulate the human proof of geometric theorems. Wu (1984) proposed a mathematical algorithm named "Wu's method" for the proof of geometry theorems from the point of view of computers. Zhang et al. subsequently improved "Wu's method" to enable automatic problem solving for nearly all geometry proof problems. Although these algorithms achieve automatic problem solving of geometry proofs, they analyze computer problem solving automatically without considering the actual problem-solving process of students. Li et al. analyzed the cognitive process of problem solving based on the cognitive model, used ACT-R to perform cognitive simulation of parallel proof geometry problems, and used the oral report method to compare and analyze the students' geometry proof process with the results of the cognitive simulation and found that the problem-solving cognitive simulation better approximated the natural process.

10.2 Cognitive Simulation of Problem-Solving Strategies

10.2.1 Methodology

The cognitive model is the basis for analyzing problem-solving strategies. The problem-solving cognitive model is constructed based on students' cognitive characteristics and subject content characteristics. The problem-solving strategy is subsequently analyzed, and the process of selecting and implementing the problem-solving strategy is described in the form of a cognitive matrix. The cognitive rigor matrix can display problem-solving strategies graphically and visually. Then, according to the content of the cognitive matrix, the cognitive process is written using a programming language (e.g., Lisp) to perform a cognitive simulation.

10.2.2 Birthday Data Collection and Statistical Strategy in the Problem-Solving Process

The Compulsory Education Mathematics Curriculum Standards (2022 edition) emphasizes the following principles: "Guide students to find and propose problems in real situations, analyze and solve problems by using observation, speculation, experiment, calculation, reasoning, verification, data analysis, intuitive imagination and other methods (Ministry of Education of the People's Republic of China, 2022)." Research has shown that the choice of problem-solving strategies is significant for students with learning disabilities in primary school mathematics when they are solving mathematical problems. Taking the application questions that primary students with learning disabilities generally reflect as examples, the difficulty lies in the choice of problem-solving strategies, i.e., how to convert the application questions into arithmetic equations (e.g., simple arithmetic equations, quadratic equations, and binary equations). After converting the questions into arithmetic equations, the students can use the arithmetic operations that they have learned to answer arithmetic questions, and the process of doing so is not complicated for the majority of students.

"Plurality" is a typical example of declarative knowledge in Unit 6, "Statistics," in the second textbook for the fifth grade. Analyzing the process of problem solving, we find that data collection is the prerequisite for solving the problem of plurality. *The Compulsory Education Mathematics Curriculum Standards* (2022 edition) stresses: "Through the language of mathematics, we can simply and precisely describe the quantitative relations and spatial forms of natural phenomena, scientific situations and daily life" (Ministry of Education of the People's Republic of China, 2022). However, when solving real-life problems, data collection is not readily available and needs to be collected and counted. Therefore, mastering data collection and statistical strategies is crucial for students to solve real-life problems.

Research on learning difficulties in students with learning disabilities in the fifth grade of primary school was selected as the research object. The analysis of the experimental data revealed that many students (e.g., DuanYC, DuanZX, etc.) had problems such as leakage and error multiplication of the statistical strategy, which led to problem-solving errors. Data processing of the "plurality" problem involves collecting the number of people who have birthdays each month and determining the month with the highest number of birthdays after counting. To reveal the data collection and statistical strategy in a more in-depth and visual way, a cognitive process was written based on the primary mathematics problem-solving cognitive model (Cui & Wei, 2012) to perform a cognitive simulation.

Figure 10.1 is a simulation used to collect the number of people who had a birthday in May (denoted by M), i.e., the number of M's. The strategist first counts the number of M in Row 1, then the number of M in Row 2, and finally the number of M in Row 3. The program output is 5, i.e., "the number of students whose birthday month is May is 5." Figure 10.2 shows a visualization of this data collection process, with the red circle indicating the last M. This visualization of problem-solving strategies can help primary school students address the issue of not being able to start processing the data. Of course, students can choose a specific problem-solving strategy according to their preferences or problem-solving habits, and the methods mentioned above are only a few of the options. Teachers can select effective teaching strategies according to the simulation results and students' actual situation in the classroom, which will help students solve problems more effectively.

According to the analysis, the declarative knowledge required for primary school students to correctly count the months with the highest number of birthdays is shown Table 10.1.

Table 10.1 shows ten ordered pairs, where p1 represents the ordered pair (1, 2), p2 represents the ordered pair (2, 3), and so on. These ordered pairs are used to compare the sizes of two numbers. Comparing the sizes of numbers has become automated for adults. If primary school children do not master these ordered pairs, they will make mistakes when comparing the sizes of numbers; i.e., they will not be able to choose the month with the highest number of birthdays.

The number of birthdays in each month is now known to be 3 in January, 2 in February, 4 in March, 3 in April, 6 in May, 4 in June, 2 in July, 3 in August, 2 in September, 4 in October, 4 in November, and 5 in December. Figure 10.3 shows a cognitive simulation of the month with the highest number of birthdays, i.e., May.

10.2.3 Interviews with Teachers

WangC's math teacher was interviewed during the study to understand WangC's usual math learning performance. The teacher's feedback is as follows.

(1) WangC is intelligent, outgoing, confident, and active. However, he needs to improve in language learning, which affects his understanding of math problems. When he was in the lower grades, he received less attention from teachers, and his language comprehension was poor. Some topical meanings of application

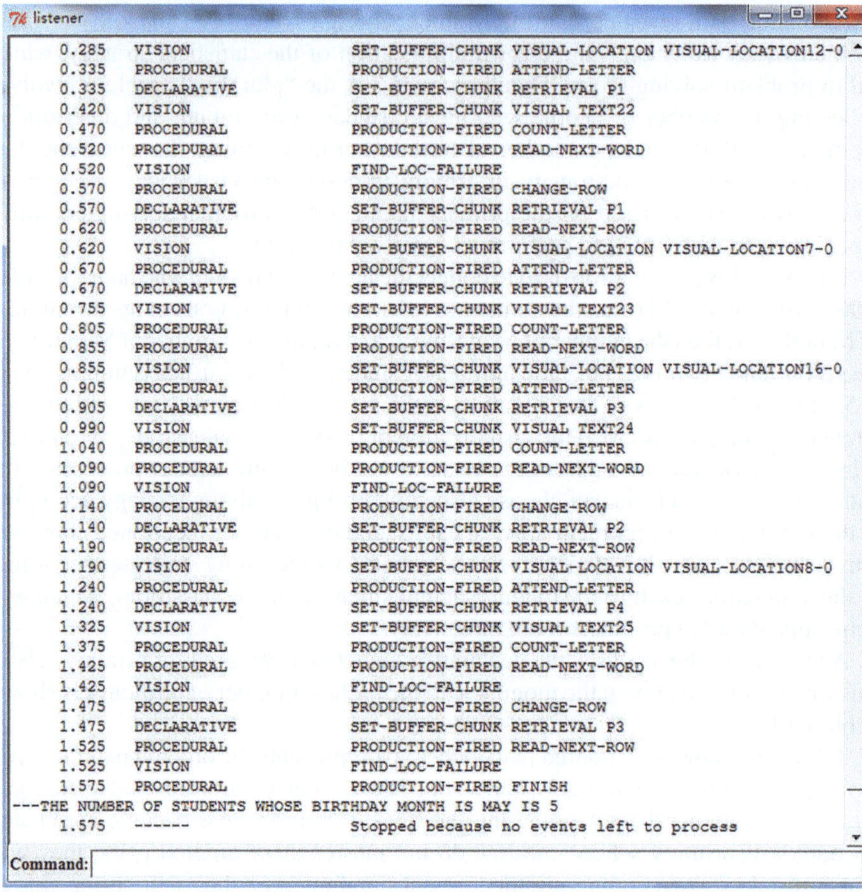

Fig. 10.1 Cognitive simulation of the data collection process

questions are not sufficiently understood, resulting in mistakes. WangC has no problem understanding what has been taught in class. As long as he can understand the meaning of the topic, he has no problem expressing the equations, but carelessness can sometimes lead to calculation errors.

(2) WangC solves problems very quickly and is among the top two students in terms of problem-solving speed. He does not have a habit of checking his work after finishing and thus sometimes has a higher error rate than other students do.

The above reflections from the student's teacher are generally consistent with the conclusions drawn from analyzing WangC's oral report data on the problem-solving process.

Fig. 10.2 Visualization of the data collection process

Table 10.1 Declarative knowledge required to count the months with the highest number of birthday celebrations

(p1 ISA count-order first 1 second 2)
(p2 ISA count-order first 2 second 3)
(p3 ISA count-order first 3 second 4)
(p4 ISA count-order first 4 second 5)
(p5 ISA count-order first 5 second 6)
(p6 ISA count-order first 6 second 7)
(p7 ISA count-order first 8 second 9)
(p8 ISA count-order first 9 second 10)
(p9 ISA count-order first 10 second 11)
(p10 ISA count-order first 11 second 12)

10.3 Implications for Teaching Mathematics

10.3.1 Helping Students Develop Good Problem-Solving Strategies

Good problem-solving strategies are essential to ensure students can answer math problems correctly. There are often multiple problem-solving strategies for the same math problem, and they often have commonalities. Teachers should encourage students to adopt multiple strategies and methods rather than "standardized" and "single" problem-solving methods to solve a problem. Good problem-solving strategies

```
listener
     1.500   PROCEDURAL          CONFLICT-RESOLUTION
     1.550   PROCEDURAL          PRODUCTION-FIRED START-COMPARE
     1.550   PROCEDURAL          CLEAR-BUFFER RETRIEVAL
     1.550   DECLARATIVE         START-RETRIEVAL
     1.550   PROCEDURAL          CONFLICT-RESOLUTION
     1.600   DECLARATIVE         RETRIEVAL-FAILURE
     1.600   PROCEDURAL          CONFLICT-RESOLUTION
     1.650   PROCEDURAL          PRODUCTION-FIRED FAIL
     1.650   PROCEDURAL          CLEAR-BUFFER RETRIEVAL
     1.650   DECLARATIVE         START-RETRIEVAL
     1.650   PROCEDURAL          CONFLICT-RESOLUTION
     1.700   DECLARATIVE         RETRIEVED-CHUNK P6
     1.700   DECLARATIVE         SET-BUFFER-CHUNK RETRIEVAL P6
     1.700   PROCEDURAL          CONFLICT-RESOLUTION
     1.750   PROCEDURAL          PRODUCTION-FIRED PERPARE-SECOND-NUMBER
     1.750   PROCEDURAL          CLEAR-BUFFER RETRIEVAL
     1.750   DECLARATIVE         START-RETRIEVAL
     1.750   PROCEDURAL          CONFLICT-RESOLUTION
     1.800   DECLARATIVE         RETRIEVED-CHUNK M7
     1.800   DECLARATIVE         SET-BUFFER-CHUNK RETRIEVAL M7
     1.800   PROCEDURAL          CONFLICT-RESOLUTION
     1.850   PROCEDURAL          PRODUCTION-FIRED START-COMPARE
     1.850   PROCEDURAL          CLEAR-BUFFER RETRIEVAL
     1.850   DECLARATIVE         START-RETRIEVAL
     1.850   PROCEDURAL          CONFLICT-RESOLUTION
     1.900   DECLARATIVE         RETRIEVAL-FAILURE
     1.900   PROCEDURAL          CONFLICT-RESOLUTION
     1.950   PROCEDURAL          PRODUCTION-FIRED FAIL
     1.950   PROCEDURAL          CLEAR-BUFFER RETRIEVAL
     1.950   DECLARATIVE         START-RETRIEVAL
     1.950   PROCEDURAL          CONFLICT-RESOLUTION
2.000   DECLARATIVE         RETRIEVAL-FAILURE
     2.000   PROCEDURAL          CONFLICT-RESOLUTION
     2.050   PROCEDURAL          PRODUCTION-FIRED FINISH-COMPARE
     2.050   PROCEDURAL          CONFLICT-RESOLUTION
     2.100   PROCEDURAL          PRODUCTION-FIRED STOP
----THE MODE IS 5
     2.100   PROCEDURAL          CLEAR-BUFFER GOAL
     2.100   PROCEDURAL          CONFLICT-RESOLUTION
     2.100   ------              Stopped because no events left to process
Command:
```

Fig. 10.3 Cognitive simulation of the month with the highest number of birthdays (partial)

not only help students, especially students with learning disabilities, successfully solve problems but also constitute meaningful ways to develop students' thinking skills in mathematics.

10.3.2 Attention to Differences in Students' Abilities

The "one-on-one" cognitive diagnosis can determine each student's mathematical problem-solving deficiencies and, in response to the students' problem-solving errors, recommend different levels of problems for students who exhibit different levels of ability. When students encounter difficulties in problem solving, teachers can provide targeted questions to gradually guide students to correctly answer the questions independently, meet the needs of students with different levels of ability, and promote the sustainable development of students' mathematical ability.

10.3.3 Early Identification of and Intervention in Cognitive Disorders in Mathematics

Difficulties are identified in advance through "one-on-one" cognitive diagnosis, and remedial measures are taken to reduce or eliminate them. By analyzing the situations of students with learning disabilities in mathematics, we can identify the cognitive barriers that lead to learning difficulties and analyze the causes of these barriers according to different types of problems and grade levels. As the saying goes, "Rome was not built in a day." Moreover, cognitive obstacles in the higher grades may have gradually developed in the lower grades. Therefore, prevention and intervention in the lower grades to address the causes of cognitive obstacles will positively impact learning in the higher grades. For example, students often make mistakes when multiplying two-digit numbers. A careful analysis of the calculation process reveals that students have already mastered the rules of simple multiplication. However, they make "digital carry" mistakes when adding two numbers; i.e., a particular cognitive pattern has been formed in the brain. When first-grade students first encounter addition, "numerical arithmetic" calculation, and problem solving, it is necessary for teachers to focus on the use of students' cognitive characteristics to explain the problem-solving process in detail, which can effectively prevent problems that may occur later in learning and effectively "obtain twice the result with half the effort."

10.3.4 Targeted Implementation of Special Counseling for Students with Cognitive Impairment in Mathematics

An analysis of the cognitive disabilities of students with learning disabilities reveals that the same or similar cognitive disabilities exist at the same grade level. An in-depth analysis of a typical cognitive impairment will be carried out to determine the reasons for its existence and formulate effective intervention measures to provide targeted counseling to students with that impairment. Changing the status quo according to which teachers explain all problems to all students in the class can save students' learning time and increase their interest in math. On the one hand, this change can provide in-depth explanations for specific math problems. On the other hand, it can truly offer an "antidote against the disease." For example, "one-on-one" cognitive diagnosis is a good choice in schools with students with learning disabilities.

10.3.5 Rational Use of Students' "Nearest Development Area" to Promote Cognitive Development

Teachers can be informed of students' cognitive level through diagnosis, develop a series of intervention measures according to the requirements of educational objectives, present them to students in a particular order and with specific requirements, and consciously participate in and intervene in students' learning process. Students acquire knowledge, skills, and problem-solving strategies, which are internalized in their original cognitive structure and contribute to forming a new structure, thus promoting children's cognitive development. Feuerstein noted that the acquisition of intermediary experience is accompanied by the process of growth in each individual and directly affects the individual's cognitive development.

10.3.6 Taking Advantage of New Technologies to Improve the Intelligence of Diagnosis and Intervention

Students with learning disabilities are a "vulnerable group" in the classroom. To promote these students' learning performance, teachers must spend more time and energy, which is currently one of the problems faced by primary and secondary school teachers. The existing computer adaptive diagnosis systems cannot meet these students' need for personalized learning. At present, the continuous emergence of technologies such as learning analytics, gesture-based technology, and virtual reality technology is gradually being applied to the field of education, which gives full play to the advantages of emerging technologies from the actual starting point of students. Combined with subject content knowledge, design, and development to meet the personalized learning needs of students' teaching robots or intelligent cognitive diagnosis and intervention systems, the automatic diagnosis of learning barriers and effective intervention measures can stimulate the automatic diagnosis of learning disabilities and provide effective interventions to stimulate students' interest in learning, improve diagnostic effects and enrich the classroom teaching environment.

References

Anderson, J. R., Fincham, J. M., Qin, Y. L., & Stocco, A. (2008). A central circuit of the mind. *Trends in Cognitive Sciences, 12*(4), 136–143.

Bai, Q. M., Liu, Y., & Qu, K. J. (2020). Research on the strategy choice of mathematics learning difficulties students in solving subtraction problems. *China Special Education, 10*, 37–43.

Bateman, B., & Kirk, S. A. (1962). Diagnosis and remediation of learning disabilities. *Exceptional Children, 29*, 73–78.

Chalfant, J. C., & Kirk, S. A. (1984). *Academic and developmental learning disabilities*. Love Pub Co.

Cui, G. Z., & Wei, X. F. (2012). Research on the cognitive problem-solving model in primary school mathematics. *Research on E-Chemical Education, 11*, 79–86.

Cui, G. Z., & Wei, X. F. (2013). Cognitive simulation and analysis of problem-solving in primary school mathematics and its pedagogical revelation—Take the problem of "addition of dissimilar denominators" as an example. *Research on E-Chemical Education, 11*, 115–120.

Geary, D. C., Hoard, M. K., & Hamson, C. O. (1999). Numerical and arithmetical cognition: Patterns of functions and deficits in children at risk for a mathematical disability. *Journal of Experimental Child Psychology, 74*(3), 213–239.

Gelernter, H., Hansen, J. R., & Loveland, D. W. (1960). Empirical explorations of the geometry theorem machine. In *Proceedings of the Western Joint IRE-AIEE-ACM Computer Conference* (pp. 143–150).

Liu, D. (2018). Research on the representation process of word problems in children with different subtypes of mathematics learning difficulties. *Psychological and Behavioral Research, 16*(4), 497–504.

Liu, D., & Mao, Y. W. (2021). A meta-analysis of the effect of intervention on mathematics skills for students with mathematics learning difficulties. *Chinese Journal of Special Education, 1*, 66–74.

Ministry of Education of the People's Republic of China. (2022). *The compulsory education mathematics curriculum standards* (2022 ed.). Beijing Normal University Press.

Tournaki, N. (2003). The differential effects of teaching addition through strategy instruction versus drill and practice to students with and without learning disabilities. *Journal of Learning Disabilities, 36*(5), 449–458.

Wei, X. F., & Cui, G. Z. (2012). Primary school mathematics problem solving cognitive model. *Electrochemical Education Research*, 11.

Wu, W. J. (1984). Basic principles of mechanical theorem proving in elementary geometries. *Journal of Systems Sciences & Mathematical Sciences, 4*(3), 207–235.

Correction to: Artificial Intelligence in Mathematics Education

Correction to:
X. Wei, *Artificial Intelligence in Mathematics Education*,
Smart Computing and Intelligence,
https://doi.org/10.1007/978-981-95-1405-2

The original version of this book was inadvertently published without the Funding Information in the copyright page of Frontmatter, which has now been included. The book has been updated with the change.

The updated version of the book can be found at
https://doi.org/10.1007/978-981-95-1405-2

X. Wei, *Artificial Intelligence in Mathematics Education*, Smart Computing and Intelligence, https://doi.org/10.1007/978-981-95-1405-2_11

Conclusion

This book is based on the research results of mathematical pedagogy and the learning sciences with respect to the process of problem solving. It takes primary school mathematics problems as the research objects to explore methods for problem solving via cognitive simulation. The research process used problem-solving cognitive process analysis, simulation and application as the main foci and included the following three parts:

(1) A cognitive model of primary school mathematics problem solving was constructed, and a method for establishing an ACT-R model of elementary mathematics problem solving was proposed. With the Polya mathematical problem-solving model, each stage of problem solving was described. Based on the thinking characteristics of primary school children, typical problems in primary school mathematics were analyzed. Based on research from cognitive psychology, brain science and cognitive neuroscience, a cognitive model of primary school mathematics problem solving is constructed. The characteristics, application scope and educational significance of CMMPS lay the foundation for analyzing the problem-solving cognitive process.

(2) Based on the cognitive model, an ACT-R model of typical mathematical problems in primary school was constructed, and a cognitive simulation of mathematical problem solving in primary school was implemented. The problem-solving cognitive simulation was conducted via the oral reporting method. According to the constructed cognitive model CMMPS, the problem-solving cognitive process of two kinds of typical problems of elementary mathematics were analyzed, namely, the "mode" and "addition with different denominators," and used the Lisp programming language to write the cognitive program to perform the ACT-R simulation. Six students in the fifth and sixth grades of a primary school were tested by the oral reporting method. Coding and analysis of the oral report data and comparisons with cognitive simulations revealed that the two were consistent.

X. Wei, *Artificial Intelligence in Mathematics Education*, Smart Computing and Intelligence, https://doi.org/10.1007/978-981-95-1405-2

(3) The application of cognitive analysis and simulation in mathematics teaching is introduced. First, the design basis and principles of the mathematical inquiry problem are proposed based on the cognitive process, and the design process and method are presented. On this basis, a typical inquiry question is designed for all knowledge points of fourth-grade mathematics. The empirical research results for classroom inquiry teaching showed that students' mathematical reasoning ability, especially those with poor mathematical reasoning ability, improved significantly after the use of typical inquiry questions.

The method and process of "one-on-one" cognitive diagnosis of the ACT-R model, which is based on an elementary school mathematics problem, were subsequently proposed, and the interaction between students who had difficulty in mathematics learning and mathematics classrooms was analyzed. The "one-on-one" method is a teaching and diagnostic method that emphasizes the process of learning; meets the different ability levels of children and their future levels of development; is combined with a dynamic assessment; considers the assessment of learning outcomes and learning process analysis; combines the evaluation functions of comprehensive identification and classification, diagnosis and prescription; and provides timely and appropriate feedback to the students in the performance of the process. Students are guided to gradually solve the problem to achieve the goal. During the experiment, the typical problems of "mode" and "cylindrical flank area" were designed, and 118 typical oral test questions for 28 students in the fifth grade and 146 oral test reports for 50 students in the sixth grade were recorded. The characteristics of the cognitive process at each stage of problem solving and the typical student's cognitive process at different stages of problem solving were recorded and analyzed. The results revealed that the effects of diagnosis and intervention were significant and then their significance for mathematics teaching were expounded.

Finally, based on an analysis of the existing methods of classroom interaction analysis, in this book we proposed a framework for classroom cognitive interaction analysis and selected the typical classroom interaction sequence in "from the perspective of different directions" in middle school (seventh grade) mathematics textbook for cognitive simulation. According to the analysis and simulation results of classroom interaction cognition, three suggestions for classroom teaching were proposed to help teachers design more effective teaching methods.

In summary, the core of this book focuses on the construction of a cognitive model for solving mathematical problems in primary school. The main contribution is to prove the effectiveness of cognitive models using computer simulation and oral report experiments. Using "one-on-one" diagnosis and intervention teaching practices, a cognitive model was applied in mathematics teaching. The analysis of the problem-solving cognitive process helps reveal the learning process and ensures effective learning.

A comprehensive analysis of the work presented in this book and the main innovations are as follows:

(1) Using the Polya mathematical problem-solving model, we refine the problem-solving phase, construct a cognitive model of elementary school mathematics

problem solving, and propose an analytical method for solving the ACT-R model in the context of elementary mathematics.

Polya presented a stage model for mathematical problem solving, which is suitable for all mathematical problems, but the internal processes of each stage were not discussed. To address this shortcoming, this research comprehensively used the research results of cognitive psychology, brain science and cognitive neuroscience to further refine Polya's mathematical problem-solving model and constructed a cognitive model of problem solving for primary school mathematics to analyze problem-solving process. This model involves mapping the cognitive process, writing a cognitive program and performing the simulation in ACT-R. The results of the simulation were consistent with the results of the oral report data.

(2) Based on the ACT-R model of typical problems in primary school mathematics, the "one-on-one" approach was used to diagnose and conduct interventions for math problem-solving students, resulting in accurate diagnoses and significant intervention effects.

Given that learning evaluations emphasize results and ignore the learning process, this research studied a cognitive model, which is based on an analysis of the cognitive process of mathematical problem solving in primary school. The research designed typical problems, recorded students' problem-solving process, solved difficult internal processes, provided targeted hints to intervene and allowed students to correctly answer the questions themselves. The diagnoses were correct. The intervention effects were significant. To a certain extent, this approach could help students with learning difficulties develop good math thinking and problem-solving habits and cultivate their interest in math learning.

In this book, the cognitive process of problem solving was explored. Despite its achievements, this research has the following limitations:

(1) In the cognitive simulation, although the knowledge points are representative, the number of knowledge points used is limited, and the number of oral reports is likewise limited.
(2) In the "one-on-one" cognitive diagnosis and intervention experiment, only the typical "mode" and "cylindrical flank area" problems were examined. Although the knowledge points were representative, the research scope needs to be expanded.
(3) In the "one-on-one" cognitive diagnosis and intervention experiment, the changes in the cognitive process of students' problem solving in different stages were analyzed.

The analysis of the cognitive process of primary school mathematics problem solving is a systematic, long-term work. This book focused on the cognitive model construction, simulation and experiment and teaching application of exploratory work. The next step will be to expand the coverage of knowledge points to increase the number of oral reports captured over a longer period to examine changes in students' cognitive processes.

MIX
Papier aus verantwortungsvollen Quellen
Paper from responsible sources
FSC® C105338

If you have any concerns about our products,
you can contact us on
ProductSafety@springernature.com

In case Publisher is established outside the EU,
the EU authorized representative is:
Springer Nature Customer Service Center GmbH
Europaplatz 3, 69115 Heidelberg, Germany

Printed by Libri Plureos GmbH
in Hamburg, Germany